# Transport and the environment

Clifford Sharp

and

Tony Jennings

Leicester University Press

1976

First published in 1976 by Leicester University Press
Distributed in North America by Humanities Press Inc., New Jersey

Designed by Arthur Lockwood
Set in Monotype Times and Monotype Gill Sans
Printed in Great Britain by
Western Printing Services Ltd, Bristol
and Bound at The Pitman Press, Bath

ISBN 0 7185 1133 6

# Contents

# Tables and figures

## Tables

## Figures

# Preface

In nineteenth-century Britain many people were able to dispel their concern about the pollution of the environment because they believed in the necessary association between muck and money. British prosperity depended upon uninhibited economic activity and any 'muck' which was created had to be accepted as a necessary evil. In our times the 'muck and money' philosophy has sustained a double attack. It has been shown that although most economic activity may inevitably cause some harm to the environment the amount of pollution can be varied. Improved technologies can be developed which create less 'muck' per unit of 'money' produced. The second, more fundamental, attack has come from those who would be prepared to sacrifice some of the 'money' in order to reduce the production of 'muck'. In recent years there has been a growing belief that the quality of life is not determined only by the quantity of goods and services available and that there may be a point beyond which further rapid increases in the size of the national income will reduce rather than augment human happiness. In pursuing some aims, such as a 'higher standard of living', greater personal mobility, and the freedom to live where they like, men have found that the achievement of other objectives, such as that of living in a quiet and unpolluted environment, has been made more difficult. This study deals with the problem of the social costs, or unwanted by-products, produced by transport services.

# Preface

We have encountered several difficulties in writing this book. One basic problem in considering possible measures for reducing the environmental costs of transport is that of balancing the value of the benefits against the costs of the remedies. A second fundamental problem relates to the development of suitable pollution-reducing technologies. We have been mainly concerned with the former, economic, problem but we have also attempted to describe some of the possible technological improvements about which we have been able to obtain information.

A major difficulty has been the scope of the study. A general discussion of broad issues would not be very useful, whereas a detailed analysis of one small aspect of the environmental problem would obviously be of limited interest. A detailed discussion of all the environmental problems of all forms of transport would not only have been beyond our capabilities but would also involve writing a study of inordinate length. The many volumes of the Roskill Report, which deals with the single issue of the siting of London's third airport, indicate the kind of enterprise which would have been required. In this study we have been doubly selective. We have written relatively little about air transport and nothing about shipping, and in discussing the environmental problems of road and rail transport we have illustrated general issues by considering specific examples. In particular we have relied on research in which we have been directly involved on such issues as the delays to other road users caused by slow-moving traffic; the environmental costs of heavy road goods vehicles; and environmental problems associated with town-centre shopping and business areas. In the economic analysis we have done our best to avoid using economic jargon even where this might have been more concise and rigorous. We have used the phrase 'vehicle mileage' to describe the distances run by road behicles even though many figures are given in kilometres, as we found the term 'vehicle kilometrage' unattractive. Chapter Five was written by Tony Jennings and Chapter Ten was a joint product. The remaining Chapters were written by Clifford Sharp. We would like to thank Mrs Joan Raby for her patience and skill in typing the manuscript. She was not only able to interpret the original handwriting but in several instances detected errors in our presentation.

Chapter one

# The theoretical background

## 1 The problem

Transport is essential to the life of mankind. As society and economic organization have become more complex so the need for transport has expanded and men have found it essential to use other forms of transport service beyond those which their own legs could supply. Modern society is entirely dependent on the provision of an abundant supply of services for transporting both goods and people. But along with the output of essential transport services certain unwanted disbenefits, such as noise and pollution of the air and the sea, are also produced. The basic question which is raised by those who are concerned about the environmental costs of transport is whether enough attention has been given to the problem of these disbenefits. It is the discussion of this question which is the central theme of this book. Although a very substantial provision of transport services may be necessary if life as we know it today is to continue, this does not mean that the present level of demand for transport is necessarily justified. It might be that some reduction in the movement of men and goods would bring about gains through the reduction of environmental disbenefits which proved to be of greater value than any losses caused by the need to reorganize economic or social activity. A probably more important possibility is that we might, if we tried, be able to provide existing transport services with significantly lower environmental costs than we do at present.

In this chapter some preliminary, but very important, issues are discussed. The basic philosophic and theoretical assumptions of the rest of the book are, as far as possible, made explicit. The particular issues discussed include the relevance of economic analysis to environmental problems; the question of how the welfare of the community should be measured; the relationship between environmental costs and 'externalities' in economic theory; the problem of compensation; the use of the pricing mechanism and government regulations as means of reducing environmental disbenefits; the nature of the environmental costs of transport; and problems of defining the environment. No attempt is made to review the literature on welfare theory and externalities, but references are included to some major sources.

## 2 Economists and the environment

Some people believe that the improvement of the environment involves issues into which the economist should not be allowed to intrude. It is sometimes argued that environmental costs *must* be reduced and economic considerations are irrelevant. Current noise levels or accident rates are held to be intolerable and the only problem to be faced is that of finding out how to make the necessary changes in technology to bring about improvements. This claim (which has been made to the authors by ardent environmentalists) implies that environmental benefits are infinitely valuable. It represents an extreme over-reaction from the situation in which environmental costs were undervalued or ignored. Most environmental benefits can only be obtained by sacrificing some other good or service (or at the expense of an alternative environmental benefit) and there is a problem of choice which makes economic considerations relevant. Some environmental improvement groups pursue their aims with an almost religious fervour and it is quite understandable that they should be suspicious of economists who try to attach price tags to environmental objectives. But in this crowded world quietness, beautiful scenery, fresh air and unpolluted seas are all becoming scarce resources which can only be obtained by giving up some alternative forms of benefit. Environmentalists may seek to persuade the rest of the community that, for example, the values placed on 'quietness' and dishwashers should be revised so that quietness is rated relatively more highly, but they are not entitled to assume that environmental improvements must be made regardless of cost. As Walters has pointed out, it is not really possible to regard 'residential quiet' as a fundamental human right which cannot, or should not, be traded.[1]

Environmental benefits should not be confused with spiritual values. We may believe that twentieth-century man is too materialistic and has forgotten that he does not live by bread alone. We may also deplore the economic hypochondria with which Western civilization sometimes appears to be afflicted and stress the importance of neglected spiritual values. But the welfare of the soul is related to the quality of the environment only in the same way as it is associated with the size of the national income. If man's experience of the world is to be divided between the spiritual and the material, then environmental benefits must be grouped with other material ends. The provision of beautiful country views, or of air freed from unpleasant smells, is an economic activity. It is just as materialistic to ask for unpolluted rivers or oil-free beaches as it is to demand fresh bread or well-printed books. The peace which passes understanding is in no way related to noise levels and the absence of motorways is neither a sufficient nor even a necessary condition for a love-filled heart.

The majority of people in the community may believe, as perhaps they do in Britain today, that the environmental disbenefits of transport should be reduced. But the problem of finding the 'right' level of environmental costs is extremely difficult to solve. The job of the economist is to discover as much as possible about the nature of the 'trade-off' which can be made between environmental goods (or the reduction of environmental bads) and other aims and to try to measure what value people place on the environmental benefits. Suppose, for example, that it was possible to reduce the noise made by all vehicles delivering to shops by 10 per cent but that the extra transport costs would increase the price of a given bag of groceries from £2 to £2·10. No civil servant or politician or member of an environmental pressure group can know intuitively whether the community would gain or lose if the noise level was reduced and the price of groceries raised. A recent report from the Noise Advisory Council concluded that 'The Panel thinks, however, that with the support and encouragement of central and local government, much can be done to ensure that Britain in 1980 is a quieter place to live in than it was in 1970. The cost of this will be high. The Panel has no doubt that the improvement which this would bring to the lives of millions of people will make the price worth paying.'[2] If the Panel did not know either the level of the costs or the value of the benefits then this statement can only be described as one of blind faith. It gives no guidance at all about how much extra quietness should be bought for the people of Britain. Any policy might be justified, from a slight tightening of the noise regulations for road vehicles to a total ban on the use of the internal combustion engine and the closure of all airports.

Some economists would argue that a distinction should be drawn between environmental disbenefits from transport which endanger 'life and limb', and those which cause nuisance, although inevitably there will be large grey areas. Given the significant measurement problems involved in placing a monetary valuation on life, it is suggested that life cannot be traded, and in determining policy measures for environmental 'life and limb' effects a cost-effectiveness approach should be used, whereby benefits are taken as given, and the least-cost method of achieving, for example, a given reduction in accident fatalities, is adopted. Other economists would argue that even if the environmental benefit is a reduction in the rate of fatal accidents there is still a trade-off situation. It would be possible to reduce the number of deaths on the road to a very low level if we agreed to have a speed limit of 5 mph. But, in effect, the community has decided that the benefits of being able to travel at higher speeds outweigh the disbenefits of the deaths and injury accidents caused. Unless life is regarded as infinitely valuable (in which case we would have very low speed limits, or ban cars from the roads altogether), the use of money values is necessary. The money value of life seeks to express, in terms of a convenient common currency, the maximum costs which the community is prepared to accept (in the form of such things as slower journey speeds or increased expenditure on street lighting) in order to reduce the accident rate. It may be argued that the money values placed on reducing noise or saving lives are too low; it cannot be logically argued that they are irrelevant.

## 3 Basis for the measurement of gains to the community

Reductions in environmental disbenefits (bought at the cost of losses elsewhere in the economy) are justified if there is a net benefit to the community. This is a fairly obvious value judgment, but difficulties arise in deciding what exactly is meant by 'the good of the community' and how it can be measured. It is assumed here that 'the community' means the people of the United Kingdom and that the welfare of each individual is of equal importance. But the utility or welfare which people receive from economic activity is largely a psychological phenomenon and there is no direct way of measuring degrees of satisfaction or wellbeing. We cannot get directly 'inside people's heads' but have to make inferences about their preferences by observing their behaviour. The usual way in which we go about this is to observe how much money people are prepared to pay to enjoy a benefit or to avoid some disbenefit. There may also be circumstances in which direct 'political' voting, for or against a proposed project or policy, is possible.[3]

There are three major problems involved in trying to use what people are willing to pay to reduce a disbenefit as a measure of the satisfaction which they will receive if the reductions take place. First, environmental disbenefits are usually public 'bads' which must be reduced for everybody or nobody. If planes become less noisy or the quantity of pollutants in car exhausts is reduced it is not possible to exclude anyone from the benefits. If quietness or unpolluted air cannot be sold directly to individuals then clearly we cannot observe how much they would pay for them. 'Willingness to pay' can only be observed indirectly by, for example, attempting to compare the price of houses in noisy areas with those of identical houses in quieter situations. Second, even if the 'benefits of reducing a disbenefit' can be sold directly it will not be possible to charge different prices to different individuals, so that some people would have been willing to pay more than the actual price with which they are faced. This is an aspect of the well-known problem of consumer surplus. Third, money is not evenly distributed, so that some individuals may have a high 'willingness to pay' not because they have an above average dislike of the disbenefit concerned but because of their extra supply of money votes. The chief disadvantage of deciding issues, such as whether to introduce legislation to raise environmental standards, by direct voting is that no account is taken of the level of likes or dislikes. The vote of a man who has a very slight preference for Situation A will count as equal to that of the man who would make almost any sacrifice to bring about Situation B.

Some of the problems of measuring 'the good of the community' can be illustrated by an example. Suppose that it was shown that the traffic noise in a particular shopping area could be reduced by 5 dB(A) but that this would add 10p to the price of an average family's weekly grocery bill. If a poll was then held amongst the 10,000 housewives using the shopping centre asking them whether they preferred the lower noise level or cheaper groceries, and the result was:

'Low noise' = 4,000
'Cheaper groceries' = 6,000

then the project to reduce traffic noise would presumably be rejected. But there are two reasons for believing that this could be the wrong decision. Most people would not be able to imagine how much benefit they would gain from a 5 dB(A) drop in traffic noise, so that the voting would, in part, be based on ignorance. (In the real world this would probably lead to a large number of abstentions or 'Don't know' votes.) And, as already argued, no account would have been taken of the degree of individual preferences for the two situations. Suppose that it was possible to measure how much the shoppers would pay for the 5 dB(A) reduction in traffic noise (and assuming for the moment that the income distribution problem can be ignored) and that it was found that the

6,000 'Cheaper groceries' voters would each have paid 8p while the 4,000 'Low noise' voters would have paid 15p each. The total benefits from the change would thus amount to £1,080 in value per week while total costs would be only £1,000 per week, so that the 'Low noise' project would be worth considering. (It cannot be said to be finally justified since some alternative forms of investment might yield a larger net benefit.) If average benefits per person were measured then these would amount to 10·8p and would exceed the average costs of 10p. If the 'Low noise' situation was introduced the gains of the minority who were each paying 10p for a benefit which was worth 15p to them, could be said to offset the losses of the majority who were forced to pay 10p for a benefit which they valued at only 8p. But whether the change can be said to benefit the community still remains a matter for individual judgment.

What advice should be given to a politician who had to choose between the 'Low noise' and 'Cheaper groceries' situations? It would be more democratic to opt for 'Cheaper groceries' but the economic arithmetic seems to show that 'Low noise' represents a more efficient allocation of resources. Is a decision-maker justified in causing Smith to be worse off even if he is sure that Robinson's gain in welfare will be greater than Smith's loss? The politician will naturally be especially wary if he knows that the Smiths outnumber the Robinsons. The economic arithmetic can itself be misleading. Economic discussion often assumes that the money value of benefits, measured by 'willingness to pay', and the actual benefit are identical, whereas the money payment is only a guide to what is happening in people's minds.

Suppose that the measurements in the Noise/Groceries example have been made very carefully, that the effect of a 5 dB(A) fall in traffic noise has been demonstrated to all the shoppers, and that 8p and 15p do represent accurately the values (measured not only in money but also in terms of alternative forms of satisfaction) of the lower noise level to the two groups. Suppose also that we allow interpersonal comparisons of utility; that is we assume that 1p of benefit is the same for each individual. It still does not follow that the benefits or extra happiness of those valuing the noise reduction at 15p would outweigh the loss of those who valued it at over 8p. The annoyance of those forced to pay 10p (per week for an increase in quietness which they valued at 8p cannot be assumed to be equal to 2p. It will not be valued at less than 2p but could well have a higher money equivalent. My sense of justice may make me very angry or unhappy if I am compelled to pay 10p for a benefit which I value at a lesser sum. The grief and pain of Mrs Smith, if she is a 'Cheaper groceries' voter, may thus exceed the happiness of Mrs

Robinson, a 'Low noise' preferrer, if the 'Low noise' situation is imposed on both of them.

In theory the problem of those made worse off by a change can be overcome by paying compensation to them. In our example, the 'Low noise' preferrers could contribute 3p a head and pay 2p to each of the 'Cheap groceries' preferrers, who would then be no worse off since with the 'Low noise' situation they are obtaining the noise reduction for the 8p which they would be prepared to pay. The 'Low noise' preferrers would seem to be better off since they would be obtaining for 13p a benefit for which they would be willing to pay 15p. But there are two major difficulties which may prevent compensation being paid in the real world. It is often quite impossible to distinguish those who would gain from those who would lose after a change. Once it was known that the payment of compensation was a possibility people would have an incentive to understate the value of benefits (or to overstate the value of disbenefits) from a change. Mrs Robinson and her fellow members of the 'Low noise' preferrers group would have an incentive to put their valuation of the benefit from the 'Low noise' situation at less than 10p if they knew that this would make them receivers rather than payers of compensation. In the real world it is most likely that, in the 'Low noise' example, the best information available would be some kind of estimate of the average value of the benefits from noise reduction. In this case it would obviously be impossible to identify the individual housewives (or families) who had a lower than average valuation of benefits and who would be net losers if they were charged 10p a week to enjoy the 'Low noise' situation. But there is a theoretical as well as a practical objection to the payment of compensation. The payment of compensation is not, as sometimes appears to be assumed in the theory of welfare economics, exactly the same as making direct payment for some economic benefit. Mrs Robinson, in our example, may be prepared to pay up to 15p to enjoy her extra 5 dB(A) of quietness but she might strongly resist the suggestion that she should pay 13p when 2p of this was to be handed over to Mrs Smith, and 1p to another fellow shopper. It is one thing to be willing to pay the necessary cost of producing a benefit, and quite another to be prepared to pay what may seem to be a form of bribe. As with the situation when the 'Cheaper groceries' preferrers were forced to pay 10p for 8p worth of benefit, the arithmetic may not measure the real gains and losses in happiness for the community. Alternatively Mrs Smith may get extra utility from receiving the transfer from Mrs Robinson. This might be called 'the transfer illusion'. The unhappiness caused to the quietness preferrers by what would be seen as the injustice of the compensation transfers might outweigh the benefit of paying only

13p for 15p worth of noise reduction. Even if the compensation was paid indirectly, through the state or some other institution, this would not seem to improve the situation very much. There would still be a transfer of income from the 'quietness preferrers' to the 'grocery preferrers' while both groups were enjoying the same amount of noise reduction.

The final problem in this discussion of the concept of the 'good of the community' and of methods of measurement is the issue, which has so far been set on one side, of income distribution. If people are to be regarded as of approximately equal importance when the costs and benefits of making environmental improvements are being considered, and if the value of benefits is to be measured by 'willingness to pay' or money votes, then there is clearly a problem if some people are given more votes than others.[3, 4] To argue that there is an income distribution problem does not necessarily imply the belief that all incomes should be equal. We may believe that the nature of human beings necessitates some unevenness in income distribution or we may indeed hold that inequality is positively beneficial and that hard work and enterprise must be rewarded. But at the same time most people, even those who believe very strongly in the advantages of free enterprise and the market economy, would agree that there should be some limitations to inequality. No-one should be allowed to starve or be deprived of essential medical care. It is no great step from these arguments to the contention that where environmental improvements are to be made, generally as the result of state action of some kind, the valuation of the benefits should not be distorted by the inequalities of income distribution. Whether we do or do not want an economic system in which some people can buy a Rolls Royce while others can only just about afford a bicycle is not the immediate concern of this study. Equally we are not disputing the right of people to buy environmental improvements which can be sold on the market, as described for example in Walters' study of aircraft noise and its relationship to house prices.[1] But where the output of public bads is to be reduced we accept the value judgment that the welfare of a rich man should be no more important than that of a poor man, and that it is therefore undesirable for the evaluation of benefits to be distorted by the vagaries of income distribution. A very similar value judgment was made by the Ministry of Transport during Barbara Castle's period of office when it was decided that the same money value per hour should be given to the leisure time savings of all adults. According to Baumol and Oates in the U.S.A., 'on balance, programs for environmental improvement promote the interests of higher-income groups more than those of the poor; they may well increase the degree of inequality in the distribution of real income.'[4]

The distorting effect of inequalities in income distribution on the measurement of the value of benefit, when the objective is some general maximization of welfare, can be demonstrated by considering differences in opportunity costs. Suppose that, in our example, Mrs Robinson was the wife of a prosperous businessman while Mrs Smith was an old age pensioner, so that they had quite different budget constraints. The opportunity cost of 5 dB(A) of quietness (at a price of 10p a week) for Mrs Robinson could then be her seventh glass of sherry (per week) while for Mrs Smith it might be two hours' warmth from her electric fire. This means that 'willingness to pay' measurements may not reflect the benefits an individual receives from environmental improvements in any systematic way. Mrs Smith might be more sensitive to noise than Mrs Robinson, and, if the noise reduction was provided free, might gain more happiness (or welfare) than her richer fellow shopper. But the different opportunity costs experienced by the two ladies would lead Mrs Smith to vote for 'Cheaper groceries' and Mrs Robinson for 'Lower noise' and Mrs Smith would be a net loser if the weekly price of the noise reduction was 10p and if no 'compensation' was paid to her.

Suppose that we keep the value judgment that each man's welfare is equally important (in relation to the decisions being considered) and we assume that it was in some way possible to measure directly, and compare, the utility or satisfaction received from economic activity of individuals. Let us assume that two men, Dives and Lazarus, live in a very simple economy in which all housing is provided free by the state, that the only commodities sold (in homogeneous units) are food and clothes, and that saving is prohibited. Food and clothes both cost £10 per unit and the respective weekly incomes of Dives and Lazarus are £80 and £40 respectively. It is proposed to improve the environment by reducing the pollution of rivers (caused by textile mills) and this will cost £20 per 'unit' per week. If the change is made both Dives and Lazarus will be provided with one unit of purer river water a week for which they must pay by a compulsory tax equal to the unit cost. Dives and Lazarus

| Number of units owned | Marginal satisfaction in 'utils' | | | |
|---|---|---|---|---|
| | Food | Clothes | | Pure water |
| | A and B | A | B | A and B |
| 1 | 20 | 10 | 5 | 9 |
| 2 | 20 | 10 | 5 | – |
| 3 | 1 | 2 | 5 | – |
| 4 | 1 | 2 | 5 | – |
| 5 | 1 | 2 | 2 | – |
| 6 | 1 | 2 | 2 | – |

both have identical tastes and patterns of satisfaction which they receive from the units of goods which they acquire. Two alternative patterns of satisfaction can be examined, which can be described as Situation A and Situation B and which differ only in the satisfaction obtained from units of clothes, and these are as set out above.

Before the environmental improvement was made the pattern of consumption (for both Situation A and Situation B) and total satisfaction would be:

| | Food | Clothes | Pure water | Total satisfaction in 'utils' | |
|---|---|---|---|---|---|
| | | Units consumed per week | | A | B |
| Dives | 2 | 6 | 0 | 68 | 64 |
| Lazarus | 2 | 2 | 0 | 60 | 50 |
| | | | | 128 | 114 |

If the environmental improvement was introduced (and both men were forced to pay for and consume the 'public good' pure water) then the new positions would be:

| | Food | Clothes | Pure water | Total satisfaction in 'utils' | |
|---|---|---|---|---|---|
| | | Units consumed per week | | A | B |
| Dives | 2 | 4 | 1 | 73 | 69 |
| Lazarus | 2 | 0 | 1 | 49 | 49 |
| | | | | 112 | 118 |

In both cases Dives would gain and Lazarus (who would be compelled to go about naked when his stock of clothes wore out!) would lose from the environmental improvement but if Situation A represented the levels of satisfaction there would be a net loss resulting from the environmental improvement while with Situation B there would be an overall gain. But in the real world 'willingness to pay' measurements must be used instead of the unknown total of 'utils'. In Situation A the maximum amount of money which could be extracted from Dives for one unit of pure water would be £41 (assuming that fractions of units of clothes could be sacrificed and that the value of each increment of one whole unit was constant), while Lazarus would pay up to £9 (surrendering 0·9 units of clothes) to enjoy the environmental improvement. Thus the total benefits in money would be valued at £50 while the cost of providing 2 units of 'pure water' would be only £40. In Situation B the money value of the benefits measured by 'willingness to pay' would be £48. If average 'willingness to pay' figures were calculated then these would be £25 in Situation A and £24 in Situation B while the average cost would be £20 per unit of environmental improvement. The 'willingness to pay' measurement in this example shows a larger net benefit when the environmental improvement would reduce human happiness (Situation A) than when it would lead to an increase (Situation B). The basic difference between the 'willingness to pay' and 'satisfaction' measurements is that the former weights preferences according to the size of income.

There is no simple method of 'correcting' the measurements of benefit to allow for unequal income distribution. One possible approach is to weight the preferences of individuals so that the poor man's 'money votes' are given increased importance. In our example the weights could be made inversely proportional to income so that the amount Dives was willing to pay for a unit of environmental improvement would be given a weight of 1 while that of Lazarus would have a weight of 2. In this case the revised average value of the benefit from the improvement would be £19·7 in Situation A and £22 in Situation B, thus giving the 'correct' result that the environmental improvement would reduce total welfare in Situation A and increase it in Situation B. Unfortunately this does not mean that the weighting procedure is a satisfactory method of dealing with the income distribution problem. Our example has been simplified by assuming that Dives and Lazarus have identical tastes, which is clearly unrealistic. If the basic preferences of the two men differed, then weighting could give 'wrong' answers. Thus if the amount of satisfaction which Lazarus would receive from reduced water pollution was less than that gained by Dives (if the improvement was free), then income-weighting would overstate benefits. If Dives had Situation A preferences and Lazarus those of Situation B then the environmental improvement would reduce total 'utils' from 118 to 112 but the weighted average 'willingness to pay' benefit would be £25·7, and would be greater than the costs of £20 per unit.

If incomes were actually rather than notionally equalized, then a quite new situation would exist in which all individuals who had changed income levels would face a different set of opportunity costs. If the incomes of Dives and Lazarus were equalized then the previous positions would be reversed. Total welfare would be increased by the environmental improvement in Situation A and reduced in Situation B (the 'before' and 'after' totals of 'utils' would be 128 and 138 in Situation A and 120 and 118 in Situation B).

The conclusion of this section is mainly negative. It has been shown that 'willingness to pay' measurements, where a proposed change will be compulsory and where no compensation is paid, must be regarded as very imperfect guides to actual changes in welfare. This does not mean that the 'willingness to pay' measurement, despite its shortcomings, may not be the best one available; but it does mean that our claims about its significance must be very modest.

In *Noise and Prices* Walters argues that 'A man's taste for quiet is measured by the amount he would be willing to pay for a quiet house', but he also suggests that quiet is a luxury good with a high income elasticity of demand (i.e. the demand for quiet tends to increase rapidly as income increases).[1] It would be very surprising if a man's enjoyment of quiet always increased with his income; differences caused by differences in income (and thus in opportunity costs) and those which reflect different degrees of susceptibility to noise must be distinguished. It is quite possible for the income factor to outweigh the taste factor and for a poor noise-sensitive man to be prepared to pay less for quietness than a rich but relatively insensitive one. Walters is of course not dealing with a situation in which noise reduction is regarded as a public good but is analysing the position where freedom from aircraft noise can be 'sold' through the choice between houses near to or distant from airport runways. 'Willingness to pay' is in these circumstances a more acceptable measure since the 'income-preferrers' (who may either have severe budget constraints or be insensitive to noise, or both) are not compelled to buy quietness. Those whose 'willingness to pay' is zero would not buy any quietness because there are no other goods which they are prepared to sacrifice in order to gain extra quietness. But in situations where quietness or some other environmental quality is a public good and must be provided to everyone or to no-one, then the effect on the welfare of the income-preferrers is also relevant. If all the consumers of the increase in quietness are forced to pay for their consumption then 'willingness to pay' measurements only tell us whether welfare is increased or decreased if we can sort out the influence of the income and taste factors. If an environmental benefit is provided 'free', the costs being met in some way which does not affect the income of those who consume the benefit, then 'willingness to pay' measurements could underestimate the total increase in welfare. People with low or zero 'willingness to pay' might gain considerable happiness from the improvement. This factor could be important in considering improvements where the benefits are not confined to one country. Suppose that the noise from British aircraft operating exclusively between Britain and a poor 'developing' country could be reduced at a cost which would just about cover the amount that British citizens were prepared to pay to gain the extra quietness. The 'willingness to pay' of the 'poor' country might be zero. The British 'willingness to pay' would then, rightly, be the only factor taken into consideration in deciding whether or not to reduce noise. But the people in the poor country might obtain just as much extra happiness out of the quieter aircraft as people in Britain. After referring to empirical evidence suggesting that the income elasticity of demand for quiet is high, Walters adds that the evidence 'is also consistent with the view that noise does not much matter in poor countries'.[1] It is certainly true that governments in poor countries ought not to devote resources to reducing noise levels when their citizens would prefer the resources to be used elsewhere. But the human beings living in poor countries might be capable of being disturbed by aircraft noise and might feel that it was a matter of consequence to them if noise levels were lowered.

## 4 Compensation or willingness to pay?

The discussion so far has assumed that the cost of making an environmental improvement should in some way be related to the value of the corresponding benefits which would be received by the community. But this implicitly assumes that the environmental disbenefit already exists. An alternative approach would be to consider how much those proposing to undertake an activity which will produce an environmental disbenefit as a by-product (this group may be called 'the polluters') ought to pay to those who will suffer the disbenefit. Many people might agree with Mishan's value judgment that where (in his example) a man annoys his neighbour with the noise from his lawnmower, it would be more equitable to make him pay compensation to his neighbour for the loss of peace and quiet rather than to expect the neighbour to pay the noise-maker to buy a quieter mower.[5] In the real world, whether or not polluters have to pay for their pollution may depend on the accidents of the legal situation. You may be able to prevent me from discharging my drains into your garden but have no legal powers to stop my children annoying you by their noisiness. But the issue of who should pay the bill for environmental benefits will be discussed more fully in the next section. The question to be considered here is whether the amount which must be paid in compensation for an act of pollution is likely to be the same as the sum that the sufferers would pay to have the pollution removed. The 'who pays whom' issue is largely academic in that actual payments are unlikely to be made in the areas which we are considering but the valuation of benefits by either 'willingness to pay' or 'compensation required' measurements could be relevant to the cost-benefit analysis of an actual proposal. If there are clear-cut groups of losers and gainers from some proposed polluting activity, if the amount of pollution produced could be continually increased or decreased by small amounts, and if polluters could compensate sufferers and sufferers could bribe polluters to reduce pollution levels, then it is possible to show that the same optimum position could be reached whether legal rights existed or not.[6] But these conditions are not met in the real situations which we are considering, where there will be a difference between compensation-required and willingness-to-pay measurements. If we start from the situation that the pollution already exists then the sufferer is, in effect, poorer than he would be before the pollution occurred. His alternative cost patterns and willingness to pay for the environmental benefit will be different in the two situations.

Suppose that in the Dives and Lazarus example the original situation was one in which the rivers were unpolluted (or at the pollution level represented by one unit of improvement from the polluted situation). The position of Lazarus would then be:

| | |
|---|---|
| Food | 2 units |
| Clothes | 2 units |
| Pure water | 1 unit |

when he would not have to surrender any income for the 1 unit of unpolluted water, which he would enjoy as a right. If the unit of pure water (worth 9 utils) was removed then the compensation required to restore the 9 utils would be 4 units of clothes (assuming Situation A) and 1 of food, so that his new position would be:

| | |
|---|---|
| Food | 3 units |
| Clothes | 6 units |
| Pure water | 0 units |

The money cost of the compensation would be £50, while Lazarus's willingness to pay for 1 unit of pure water when the original situation was:

| | |
|---|---|
| Food | 2 units |
| Clothes | 2 units |
| Pure water | 0 units |

would only amount to £9.

## 5 Environmental bads and externalities

The question of responsibility for environmental disbenefits was raised in the last section. In economic analysis, environmental disbenefits are usually treated as a form of 'externality'. The exact definition of an 'externality' has been a matter of much discussion.[4,6,7] A recent definition by Meade is that 'An external economy (diseconomy) is an event which confers an appreciable benefit (disbenefit) on some person or persons who were not fully consenting parties in reaching the decision or decisions which led directly or indirectly to the event in question.[8] The environmental costs of transport activity are described in the next section, so another non-transport example is used here.

In a suburb of Leicester, there is a dye-works that produces a pervasive sickly smell which some local residents dislike. This is a clear example of an external diseconomy, affecting the welfare of people who have no control over the operations of the dye-works. Now suppose that the smell could be reduced but that this reduction would cost the dye-works £10,000 a year. The citizens of the suburb who disliked the smell could band together and find out how much they would be willing to pay in order to persuade the dye-works to make the smell-reducing improvement. If the sum was less than £10,000 a year then, in a free-bargaining situation, the improvement would not take place. This situation is what is called in economic jargon 'a potentially relevant externality'. Since the benefits of reducing the disbenefit caused by the smell would (according to the 'willingness to pay' measurement) be less than the costs, it is therefore in some sense economically 'efficient' for the dye-works to continue its activities unchecked, and for the people of the suburb to put up with the smell. There is a problem of equity in that although there would be a net gain to the whole community from the continued operation of the dye-works the people suffering from the smell would be losers. In theory they could be paid compensation which would make up for their inconvenience, but, as we have already shown, this is difficult to organize in practice.

The second possibility is that the people who suffer from the dye-works' smell might be willing to pay more than £10,000 to gain the benefits from the possible reduction of the nuisance caused by the smell. In this case there could be gains from trade between the people of the suburb and the dye-works operators. This is an example of what is called a 'Pareto-relevant' externality in economic welfare theory. The dye-works directors would accept a 'bribe' of £10,000 (or more) to reduce the smell. If they were compelled to pay compensation of £10,000 or more to the local inhabitants in order to continue producing the smell they would choose the alternative of reducing it.

The most obvious case for some kind of state action is where this second situation exists but where there are barriers which prevent people from organizing themselves to pay a 'bribe' or to demand compensation. Thus the people of the Leicester suburb cannot, in the present state of the law, compel the dye-works operators to compensate them for the smell and they might find it very difficult to agree on their respective contributions to the necessary 'bribe'. But although there is a prima-facie case for state intervention when a Pareto-relevant externality exists,[6] and it is therefore clear that the economic system is working inefficiently, it is not at all clear what action the state should take. If perfect knowledge existed then there would not be any problem. Suppose that it was known that the smell from the Leicester dye-works could be reduced from level S5 to S4 (measured on some scale which increased with the unpleasantness of the smell) for a cost of £10,000 a year while the benefit of local inhabitants from the reduction would be £15,000. In these circumstances the state could simply order the dye-works to make the reduction. If the same condition applied in all dye-works in Britain then an order could be introduced making S4 the maximum permissible level for smell 'emission'. In Britain maximum 'pollution levels' are laid down for such disbenefits as the noise created by road vehicles or the visible smoke emitted by the exhausts of lorries. But perfect knowledge of the costs and benefits of environmental improvements does not exist in any real-world situation (not even in cases like vehicle noise and smoke emission where pollution levels have been regulated by legislation). This means that legal pollution maxima may be set too high or too low, in relation to costs and benefits, and could quite easily result in a net loss in welfare.

Another possibility is for the state to put a tax on processes creating environmental disbenefits. A tax which varies with the level of pollution has the advantage that it offers firms a choice between reducing pollution levels and paying tax so that they will operate at the 'efficient' level where any further reduction in pollution level would just cost more than the corresponding tax reduction.* But the unit rate of tax could only be fixed at the 'right' level if the government knew the unit cost of the disbenefit to the community. In the absence of this knowledge the tax (like maximum pollution levels) could be set too high or too low, and is very unlikely to be 'just right'. Taxes also have other disadvantages. The producer of the disbenefit would have to pay tax on units of pollution which were 'efficient' (in the sense that their removal would cost more than the value of the benefits) and this would distort the prices of their product. The use made by the state of the tax revenue

* If T was the tax rate per unit of pollution and the cost of reducing pollution was some function C (and if pollution levels were continuously variable) then the firm would reduce pollution to the point $T = dC$ where the tax rate and the marginal cost of reducing pollution were equal.

could also bring distortions to the economy. Turvey, in an article summarizing the discussion on the divergence between social cost and private cost, argued that taxes imposed on the causes of a diseconomy would lead to complicated results and were generally unwise.[9] Baumol and Oates consider that in some circumstances taxes may be less flexible than direct controls.[4] Flexibility is important where standards may change and where mistakes may need to be corrected.

The immediate objective of pollution taxes is sometimes said to be that of 'internalizing' disbenefits. This is reflected in the popular value judgment that polluters ought to pay the cost of the pollution they cause. Road haulage operators should pay for the noise and general nuisance created by their lorries and oil companies should pay for the beaches they spoil and the fish they kill when their tankers leak into the sea. It will be argued later in this study that the external disbenefit of road wear caused by heavy vehicles can be costed with at least some degree of confidence and that there may be advantages in using a tax to 'internalize' this externality. External disbenefits can also be internalized by extending the area of operation of pollution-causing undertakings. Thus if the Leicester dye-works owned all the houses in the area affected by the dye-works' smell, and if the disbenefit of the smell was reflected in the rents which could be obtained from the houses, then the dye-works' directors would have an incentive to reduce the smell to an 'efficient' level. But such extensions of operation are rarely possible. The scale of operation required to 'internalize' many externalities would be such that, even if it could be achieved, it would not necessarily lead to an 'efficient' solution of the problem.

The analysis of externalities and Pareto-efficient solutions is based on the assumption of the existence of two distinct groups of people, pollution creators and sufferers, who may be gainers or losers in any change in the situation. But with such disbenefits of transport operations as traffic noise the gainers and the losers, the polluters and the polluted, are all in the end the same group of people – the whole community. Both goods and passenger transport are essential services consumed by the whole population of Britain. If transport operators adopt high-cost technologies in order to reduce pollution then the 'losers' will not be the transport operators, but transport consumers. If all road hauliers are in some way persuaded or compelled to use more expensive, but quieter lorries, then the costs will be passed on eventually to all those who consume goods that have been transported. It is true that the disbenefits of transport operations are not distributed evenly, and that people who have a trunk road running past their front door or a railway at the bottom of their garden may suffer more than an average amount of disbenefit. But almost everyone in the community bears some part of the environmental costs of transport operations. There is thus no possibility of removing the problem of the environmental costs of transport by 'internalizing' them and then allowing the automatic working of the price mechanism to provide an efficient solution. The costs are in a sense already indirectly internalized. Even if all transport was directly provided by the state for its citizens the problem would not disappear. Turvey has argued that when the problem of social costs cannot be solved by negotiation between the polluters and the polluted 'the theorist should be silent and call in the applied economist'.[9]

Even if the polluter-pays principle does not provide a method of reaching an optimum level of pollution it may fulfil a more modest role. The Organisation for Economic Co-operation and Development (OECD) has advocated the use of the polluter-pays principle as a means of ensuring that 'acceptable' levels of pollution are reached.[10] It is assumed that governments can somehow determine the levels to which pollution should be reduced. A tax imposed on polluters will then assure that they reduce pollution in the most efficient (or least cost) way. When the polluter-pays principle is used to achieve this limited objective the identification of those ultimately responsible for causing the pollution is not necessary and whether or not the increased cost is borne by manufacturers or passed on to consumers does not affect the applicability of the principle. Judith Marquand and D. Allan argue in a 'Note' on the OECD study that 'Whether it is the production process which gives rise to the pollution, or the consumption of the final good or service produced, the costs of the pollution abatement should be borne at whatever point is most effective and most (administratively) convenient in the chain of production and consumption services.'[10]

The revenue from a tax designed to reduce pollution to an 'acceptable' level could in theory be zero since an optimum tax would equal the marginal cost of the last desired unit of reduction in pollution, and apply only to pollution above the 'acceptable' level. In practice mistakes would be made in fixing the level of the tax and different polluters might have different unit pollution reduction costs. The justification for a tax which polluters could not wholly escape (because the cost of reducing pollution was greater than the unit rate of tax or because a pollution-reducing technology did not exist) would depend on circumstances. A tax might be used to discourage manufacturing processes which were environmentally costly if the good produced was considered to be of little importance in the economy. A tax on pollution might also be levied to produce revenue to compensate those who suffer from the pollution. But in other circumstances a tax on polluters might have little point. The extreme case would be where the whole community both benefits from consuming the goods produced and suffers from the pollution. If the costs of further reductions in pollution would exceed the benefits then nothing would be achieved by the community taxing itself in order to compensate itself. As already argued the costs and benefits of some transport operations may both be so widespread that a 'compensation' tax would be difficult to justify.

## 6 The environmental cost of transport

The discussion so far has been in general and theoretical terms and it is now necessary to describe the particular environmental costs with which this study is concerned. Some of these may be relevant to all forms of transport, others only to a specific mode. The word 'environment' does not have any very precise connotation, so we have extended the list to include all the unwanted by-products of transport operations. These disbenefits do differ in type and could be divided into subcategories. Some important disbenefits are 'subjective' in the sense that they reduce human welfare but do not use up scarce resources. Noise is a disbenefit because people dislike it, but it does not reduce the size of the measured National Income. Other disbenefits are in a sense economic because they involve the consumption of resources. Road wear is the most obvious example and a narrow definition of 'environment' could lead to the exclusion of this class of disbenefit. (But the cost of road wear is not only that of the labour and materials needed to repair the roads; people may dislike riding on bumpy roads.) Possibly in a category by themselves are accidents, which are costs that transport users largely impose on each other, and which involve the difficult problem of evaluating human life.

One social cost of transport excluded from this study is that of road congestion. This is in no sense a by-product of transport operations but is merely a symptom of a lack of co-ordination between the demand for road space and its supply. Many of the unpleasant aspects of heavy traffic, such as difficulty in crossing roads, are included in our list of environmental costs, but these costs would still be there (and might sometimes be worse) on uncongested roads. Delays caused by especially slow-moving vehicles, as opposed to those caused by the vehicle flow exceeding the road design capacity, are considered in this study.

The list of environmental costs of transport which will be discussed is therefore as follows.

1. Noise from vehicles, aeroplanes or trains.
2. Air pollution by vehicle exhausts.

3. Ground-borne vibration caused by heavy vehicles and affecting roadside buildings.
4. Accidents.
5. Delays caused to other road users by slow-moving vehicles.
6. Road wear.
7. Visual intrusion by vehicles or transport infrastructures into urban or rural landscapes.
8. The severance of communities by railways, canals, or busy roads.

**Summary**

*1. Environmental benefits do not have infinite value and it is therefore essential to try to compare the costs and benefits of any change which will reduce environmental but increase other costs. The 'cost effective' approach in which an arbitrarily determined environmental improvement is brought about at the lowest possible cost is very unsatisfactory because even the minimum cost may exceed the value of the benefits. But since it is very difficult to quantify the value of environmental improvements it may be necessary to accept a possible margin of error in the evaluation of benefits which is greater than that used in putting a money value on costs.*

*2. Where state intervention is necessary before an environmental improvement can be obtained it is assumed in this study that the objective of the state is to improve the 'general good' of the community and that, ideally, the good of every citizen should be regarded as being of equal importance.*

*3. Maximizing the general good, or social welfare, is an impossible task because the happiness or unhappiness caused to people by economic changes cannot be measured. The best that can be done is to use either the evidence of how much people will pay for a given environmental improvement as an indication of the 'happiness' it will give them, or to consult people directly about proposed changes and then count their votes. The chief advantage of the 'willingness to pay' measurement is that people have a number of 'money votes' and can thus express the intensity of their desire for a particular change. In 'political' voting where each citizen only has one vote a majority who mildly favour a change will be able to outvote a minority who oppose it very strongly (or vice versa). Other things being the same, the willingness to pay measurement is thus to be preferred (but see point 5 below).*

*4. Where some people (the polluted) are compelled to suffer a disbenefit because of the economic activities of a second and distinct group of people (the polluters), then it is difficult to dispute the value judgment that the polluters ought to compensate the polluted and that the polluted should not be compelled to bribe the polluters to reduce the level of their activity. In the abstract world of 'high theory' in economics it can be shown that the same optimum level of pollution can be reached in either case. But in the real world where income distribution effects cannot be assumed away the legal situation in relation to pollution activities is important. The actual payment of optimum levels of compensation is also rarely possible. Those who require little or no compensation for some environmental disbenefit cannot be distinguished from those who would need large money payments to be as 'well off' with the payments and some level of disbenefit as they would be without either. The payment of compensation is also not necessarily neutral in its influence on human happiness as is assumed in some theoretical discussions. A 'transfer illusion' may operate. A man who was prepared to 'buy' an environmental improvement for £1 a week might not be equally willing to pay 80p for his own share of the benefit and to contribute 20p towards his neighbour's costs.*

*5. The biggest difficulty in using the 'willingness to pay' measurement for environmental benefits is that some people have more money, and therefore more votes, than others. Our willingness to have some inequality in income distribution does not mean that all its consequences must be accepted. If no adjustments are made for income inequalities when state action to reduce environmental disbenefits is under consideration then the requirement that each person's welfare should be of equal importance is violated. Changes involving public 'bads' which are justified by crude willingness to pay measurements could nevertheless lead to a reduction in human happiness. There is no simple way of making adjustments for income distribution in cost-benefit calculations but there is a strong case for questioning the desirability of any change in which the benefits accruing to a richer minority appear to outweigh the losses of a poorer majority (or in which the average gross value of benefits is greatly affected by the high willingness to pay values of a relatively small number of beneficiaries). In other circumstances 'willingness to pay' measurements may underestimate the total increase in happiness caused by an environmental improvement. People whose willingness to pay for an environmental benefit is low or zero may nevertheless be affected by the relevant disbenefit. It is not easy to construct a weighting system which would 'correct' existing income distribution.*

*6. The environmental costs of transport discussed in this study are: noise; air pollution; vibration; accidents; delays caused by slow-moving road vehicles; road wear; visual intrusion; and the severance of communities.*

Chapter two

# The demand for transport

## 1 Introduction

The most obvious way in which the environmental costs of transport operations could be reduced would be to rearrange the economic life of the nation in such a way that less transportation was required. The purpose of this Chapter is to examine the nature of the demand for the carriage of goods and people in Britain and to consider the choices which might be open to society if a policy of limiting transport demand were to be adopted.

## 2 Factors influencing the demand for transport of road freight

There are several ways in which the demand for goods transport can be measured. The most common unit of measurement is 'ton miles' or tons of goods carried multiplied by the length of haul. Road goods transport, with which most of the rest of this discussion is concerned, can also be measured by vehicle miles, or by the size of the total fleet of lorries in the country. Vehicle mileage is probably the most useful measure of demand considered in relation to environmental impact.

If ton mileage remains constant but vehicle mileage is reduced, by, for example, an improvement in load factors, then environmental costs will fall. Similarly the existence of a larger lorry fleet will not have any significant effect on the environment so long as vehicle miles on the roads are not increased. (The regulations on maximum driving hours and vehicle miles could result in the use of a larger fleet of vehicles with lower average annual mileages). Vehicle miles is a rather crude measure of environmental cost since, of course, the environmental impact of lorries will also vary with the routes on which they operate. Ton miles, vehicle miles, and total vehicle population will all be related fairly closely but it is assumed in the following discussion that the objective is to minimize vehicle miles.

Models have been used to predict the demand for goods transport but these are not very useful for analysing the nature of this demand or for examining policies which would minimize ton miles. The predictive model developed by Tulpule[1] and Tanner[2] is based on the assumption that there is a relationship between the total ton miles of goods carried in one year and gross domestic product. If $T_i$ is total ton miles per annum and $X_i$ is GDP then

$$T_i = \alpha X_i$$

Road goods transport ($RT_i$) is some proportion p of this total

$$RT_i = p(\alpha X_i)$$

The average loads per vehicle, $\bar{L}_i$, are then estimated from simple extrapolations of past trends in average vehicle capacities, $\bar{C}_i$, and load factors, $\bar{F}_i$,

$$\bar{L}_i = \bar{C}_i \times \bar{F}_i$$

(assuming that load factor is expressed as a proportion of capacity). Vehicle miles for any year, $VM_i$, are then given by

$$VM_i = \frac{P(\alpha X_i)}{\bar{L}_i}$$

The model assumes that total ton miles are some pre-determined function ($\alpha$) of GDP and that changes in average load will follow some simple and foreseeable trend. The predictive models do not entirely ignore possible policy changes but subsume them all in a few simplified assumptions. Thus Tanner assumes that the ratio of tonne km to GDP will fall by 1 per cent a year (or that tonne km carried by all modes will increase in proportion to the two-thirds power of GDP). He also assumes that policies intended to encourage rail goods transport will result in the recent decline in rail freight (the average annual decline in tonne km from 1962 to 1972 was 1·3 per cent) changing to an average increase of 0·5 per cent per year. A further important assumption in the Tanner model is that the rate of growth in average vehicle carrying capacity will be lower in the period up to 2010 than it has been in the past 14 years, and that this rate of growth will be influenced by the rate of growth of GDP. Thus with low GDP growth he assumes that average carrying capacity will increase by 1·6 per cent per year in the period 1975–80 and by 0·7 per cent per year in the five-year period 2005–10. The corresponding figures for high GDP growth are 2·3 per cent and 1·1 per cent respectively. Tanner's model is used purely for predictive purposes and he points out that 'There is little doubt that alternative policies could have a substantial effect on the figures'.[2]

An alternative 'explanatory' model of freight transport demand, which is useless for predictive purposes but which can be used to illustrate some possible policy choices, can be developed. Suppose that there were five points sending out or receiving road goods traffic arranged as in the diagram below.

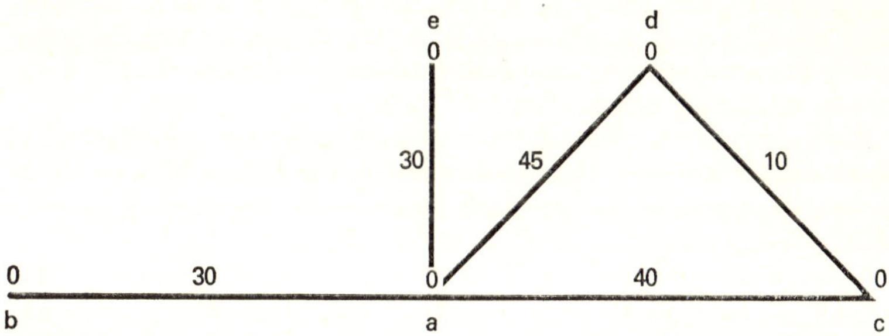


The figures show the mileages between each traffic point. Let us assume that the daily traffic flows were all of single 10-ton consignments and that the origins and destinations were:

a ———→ b
b ———→ a
a ———→ c
a ———→ d
a ———→ e

These flows represent the 'basic' demand for road goods transport. This concept is similar to that of 'standardized ton miles' developed by Bayliss (see reference 21, Chapter 9). If $T_i$ are tons to be carried and $M_i$ the appropriate distances between origins and destinations then the total daily basic ton miles are $\Sigma T_i M_i$ which in the case of our example would be 1,750 ton miles. But actual ton miles and actual vehicle miles cannot be predicted from this explanatory model. The actual vehicle mileage, ($\Sigma m_i$), if it was assumed that each consignment occupied one lorry, would be given by

$$\Sigma m_i = 2\Sigma M_i$$

(350 vehicle miles in the case of the example); but consignments can, of course, be amalgamated, or may need to be carried by more than one vehicle. If actual ton mileage was $\Sigma m_i t_i$ and the average load factor was F then vehicle mileage (VM) would be given by

$$VM = \frac{\Sigma m_i t_i}{F}$$

Thus, in the example, if a single 20-ton capacity vehicle was used on the run a–c–d–a, and 10-ton capacity vehicles on the routes a–e–a and a–b–a, then total ton mileage would equal 1,800, the average load F would be 8·37 tons and total vehicle mileage would be 215. But since, when working from theory rather than observation, it would be necessary to know the vehicle mileage before calculating the average load, the vehicle mileage equation is not very useful.

It is clear from the explanatory model, however, that the three factors which will determine vehicle mileage will be the volume of goods to be carried, the length of haul for each consignment, and the possibility of amalgamating loads.

As has already been stated this study avoids a direct discussion of the difficult problem of the relationship between economic growth and human happiness. If the rate of economic growth was deliberately reduced, then one of the benefits would be a lower demand for goods transportation and a reduction in the associated environmental costs. But there would be many other more important costs and benefits and the growth controversy will obviously not be decided in the context of transport alone. All that can usefully be said in this discussion is to point out that the environmental costs of transport are just one part of the total environmental and ecological costs of economic growth. If the community became convinced that a policy of slower (or even negative) economic growth would bring net benefits then no doubt it would be worth while to pay special attention to those industries which generate a high demand for transport (just as industries creating high levels of pollution would need examination). But at present it seems unlikely that reductions in the output of goods will be used as a means of reducing the environmental costs of transport.

The demand for goods transport (measured in ton kilometres) would be reduced if the average length of haul could be shortened. The extreme case, in the basic model, would be to move all industries and people to the one centre 'a'. The location of industry and of the population is determined by complex factors and it is obvious that any attempts to move factories and people into more dense concentrations would have many other effects besides reducing the demand for transport. A major factor affecting industrial location has been the economies of scale achieved in many industries by concentrating production in a few major plants. Scale economies generally offset the increased transport costs which result from centralizing production. According to Census of Production data the total number of manufacturing establishments in the U.K. in three recent surveys was:

1958 = 92,785
1963 = 84,747
1968 = 82,343

The output of manufacturing industry, measured at constant factor cost, during the same period was:[3]

1963 = 100
1958 = 84·4
1963 = 100·0
1968 = 121·4

The number of separate manufacturing establishments fell by 11·2 per cent between 1958 and 1968 while manufacturing output increased by 43·8 per cent during the same period. This greater concentration of manufacturing activity would not necessarily increase the average distance which goods had to be carried. Although factories are, on average, likely to have become located away from final consumers, there may have been some reductions in the distances which raw materials and semi-finished materials were carried. Thus if a car-body pressing plant was closed and the work carried out instead at the main car assembly factory the total ton kilometres of transport work would probably be reduced (unless the body-pressing plant happened to be located on the direct route between the steelworks and the final assembly factory). But the concentration of the manufacture of some consumer products may, on balance, have increased the average length of hauls, because reductions in the length of haul for raw materials are more than offset by increases in the length of haul of finished products. The concentration of manufacturing activity is also likely to increase the demand for transport where the finished product is more bulky than the raw materials. Where outputs occupy more space than inputs and the concentration of manufacture results in shorter hauls for inputs but longer hauls for outputs total ton mileage figures might remain the same. But the demand for transport would effectively be increased if average vehicle loads were reduced because the bulky manufactured goods caused load factors to deteriorate. Goods with poor loadability will use up all the volume available in a vehicle before it has a full load measured by its weight capacity.

'Bread and flour confectionery' is an example of an industry in which outputs may be more bulky than inputs and in which greater concentration of manufacture may lead to an increase in the average length of haul, or to a deterioration in load factors and average loads, or to both of these results. Table 1 shows recent trends in this industry.

**Table I Concentration in the 'bread and flour confectionery' industry: U.K.**

| Year | Number of establishments | Value of net output (£m) |
|---|---|---|
| 1958 | 2371 | N.a. |
| 1963 | 1506 | 159·8 |
| 1968 | 1046 | 208·2 |
| 1970 | 971 | 270·3 |

Source: *Census of Production*, 1958, 1963, 1968, and 1970.

The community will benefit from the concentration of manufacturing in a few larger plants when economies of scale in production offset any increase in transport costs, so long as it is the full social costs of transport that are measured. If manufacturers consider only the private costs of transport when planning the location of their plant their factories could be wrongly sited from the point of view of the community. But it seems unlikely that the concentration of manufacture will greatly expand the total demand for transport since, in most situations, increases in the average length of haul of the finished product will be offset by reductions in the average length of haul of inputs.

It is not easy to discover what has been happening to the average length of actual hauls. Data on lorry hauls is difficult to interpret as the proportion of traffic carried to intermediate warehouses or depots may change. If a chocolate manufacturer decided to distribute to shops via warehouses rather than direct from his factory, individual 'lorry hauls' might be shortened while the total length of haul could remain constant or even be increased. Figures on 'length of haul' in recent Department of the Environment surveys are obtained by dividing estimates of total annual ton kilometres by those of total tons carried. They thus represent a 'ton-weighted' average length of haul.[5] As empty mileage is ignored the Department of the Environment ton-mileage figure cannot be used directly for estimating vehicle miles. The Department of the Environment figures indicate that the 'ton-weighted' length of haul has been decreasing in recent years.

**Table 2 'Ton-weighted' average length of haul for road goods traffic**

| Unladen weight Over | Not over | Average length of haul 1967/8 | 1970 | 1971 |
|---|---|---|---|---|
| Public haulage | | miles (km) | | |
| – | 5 | 25 (40) | 27 (43) | 22 (35) |
| 5 | 8 | 42 (68) | 33 (53) | 32 (51) |
| 8 | – | 63 (101) | 57 (92) | 49 (79) |
| Total | | 37 (59) | 37 (59) | 34 (55) |
| Own account | | | | |
| – | 5 | 18 (29) | 18 (29) | 20 (32) |
| 5 | 8 | 26 (42) | 25 (40) | 26 (42) |
| 8 | – | 31 (50) | 32 (51) | 30 (48) |
| Total | | 21 (34) | 22 (35) | 24 (39) |
| All vehicles | | | | |
| – | 5 | 20 (32) | 21 (34) | 21 (34) |
| 5 | 8 | 35 (56) | 30 (48) | 30 (48) |
| 8 | – | 50 (80) | 48 (77) | 41 (66) |
| Total | | 28 (45) | 30 (48) | 29 (47) |

Source: *The Transport of Goods by Road, 1967–68* (Department of the Environment, 1971).

These figures suggest that the average 'ton-weighted' length of haul tended to remain stationary or fall during the period covered by the survey. The average length of haul of the heaviest vehicles (over 8 tons unladen weight) fell most noticeably. It seems unlikely that these figures reflect any fundamental movement of industry nearer to the areas where their products were consumed. The total figures for all vehicles show little change and it may be that heavier vehicles were taking over more short-haul work from smaller lorries. It is also possible that sections of industry which create demand for relatively short hauls of traffic (including both inputs and outputs) have grown more rapidly than those which generate demand for long-distance movements. The relative success of freightliner operation may have removed some long-distance traffic from road to rail, but this is not likely to have had a significant effect on the road haul figures during the period 1967–71. If there was any tendency for average hauls to be lengthened by the concentration of manufacturing activity this seems to have been offset by other factors.

Any policy aimed at decreasing the demand for transport by reducing the average length of the haul must involve very complex considerations. As has been argued, dispersal of manufacturing activity so that goods are made near to the main centres of demand will not necessarily reduce the total need for goods transport since shorter hauls for outputs may be offset by longer hauls for inputs. In the long run people and industry could be concentrated in a few major conurbations, and this would undoubtedly reduce the demand for both goods and passenger transport. But this would involve reversing the trends of the last 70 years and reconstructing towns as densely populated as those of Victorian Britain. This would itself involve very considerable social and environmental costs, which would almost certainly be greater than any environmental benefits resulting from the fall in the demand for transport.

It is possible that some reduction in the demand for goods transport could be brought about as a result of the development of marketing agreements between firms producing similar products. Suppose that there are two biscuit manufacturing firms, Northern Biscuits and Southern Cookies, the former with its factories in the Glasgow area, the latter having its manufacturing activities concentrated in London. If both firms market their product all over the U.K. then lorries or freightliner trains carrying biscuits from London to the North will pass other lorries or trains transporting biscuits from Glasgow to the South of Britain. It is obvious that the demand for transport generated by the biscuit industry would be reduced if Northern Biscuits supplied the whole of the market in northern Britain while Southern Cookies met the demand in the South. But there would be some obvious disadvantages if such a market-sharing agreement was made. Any benefits such as improvements in technology which resulted from competition between the two companies would be lost. If the companies now had a monopoly, or near-monopoly, in their areas monopoly pricing policies might be adopted. Consumers would also lose by having their area of choice reduced. People in the North who had preferred the products of Southern Cookies, or southerners who had a preference for biscuits from the northern company, would suffer some loss. Once again there is a complicated trade-off situation in which it is impossible to make any generalizations about whether the advantages of reductions in the demand for transport would, or would not, be offset by the value of the various disbenefits. Perhaps the situation in which a geographical division of markets would be most advantageous is in industries where there are some economies of scale which will result from concentrating manufacture in large plants; where the product is reasonably homogenous; and which generate a large demand for transport (or where transport costs are a significant proportion of total costs). Where economies of scale in manufacture are not important firms can reduce transport demand and cost by building plants in different markets. Thus, in the previous example, Southern Cookies might build a factory in the North and Northern Biscuits could manufacture in the South. (The earlier discussion about industrial concentration or dispersion is of course relevant here. Moves to supply separate markets from local factories will only reduce net transport demand so long as there is not an offsetting increase in the movement of raw materials). But if scale economies are such that only one plant is required to supply the whole of a firm's share of the home market, then regional factories are ruled out. Where rival consumer products are virtually indistinguishable, except for different packaging, and it is apparent that no single company is likely to gain the whole of the home market (even although plants are of sub-optimum size), regional marketing agreements which reduced transport demand and transport costs might be advantageous. A possible example would be in the marketing of detergent which (despite advertisements about the inferiority of Brand X) is a reasonably homogenous product, with some scale economies in manufacturing and significantly high transportation costs (7·7 per cent of the value of net output for manufacturing firms covered by the 1968 Census of Production returns). Some possible ways in which vehicle mileage might be reduced in town-centre shopping areas are discussed in Chapter Ten.

Vehicle mileage will be reduced when a number of consignments can be carried in one vehicle. Thus in the 'explanatory' model the amalgamations of consignments from 'a' to 'c' and from 'a' to 'd' would mean that one vehicle could deliver to both centres on a round trip of 95 miles. If the loads were not amalgamated, and separate trips from 'a' to 'c' and 'a' to 'd' were necessary, then the relevant vehicle mileage would increase to 170. The degree to which loads can be amalgamated depends upon several other factors. The most obvious requirement is that routes should coincide. The concentration of industry and population into major centres could result in increased opportunities for amalgamating loads as well as in reducing the basic length of haul requirements. The planning of industrial locations so that flows between major centres are approximately equal in size, and vehicles can obtain back loads, will also help to increase average loads and reduce vehicle mileage. But these are very long-term solutions, and, as already argued, may have other adverse environmental implications.

An important factor influencing the ability to amalgamate loads which is to some extent in the control of policy-makers, is the carrying capacity of lorries. In the 'explanatory' model it was assumed that vehicles would have the capacity to amalgamate loads but in the real world this may be limited by vehicle capacity. Since the maximum gross weight of lorries in the U.K. is fixed at 32 tons, and since the unladen weight of a 32-ton articulated lorry will be at least 10 tons, the effective maximum capacity of road vehicles in this country is 20–22 tons. Suppose that there was a daily two-way flow of traffic betweeh two centres 'a' and 'b', 100 miles apart, of 2,000 tons each way, and that this traffic was in every way suitable for amalgamation, then it could be carried in 100 20-ton capacity vehicles running a total of 20,000 vehicle miles. But if 26-ton capacity vehicles were operated, only 77 lorries would be required and the total daily vehicle mileage would fall to 15,400. Actual average vehicle loads will be affected by all the other factors determining the possibility of amalgamation, but maximum vehicle capacities will have an important influence.

The possibility of amalgamation will be affected by the nature of the commodities which are to be carried. There are many goods which cannot be carried together in the same vehicle. Foodstuffs and cement or petroleum and furniture are obvious examples. A survey which examined the possibility of using an interchange depot for lorries delivering goods to shops in Chichester showed that about 50 per cent of the goods carried in retail distribution could not conveniently be amalgamated.[7] Amalgamation is also affected by the loadability of goods. Poor-loadability goods, which are bulky in relation to their weight, may use up the physical capacity of a vehicle before the weight limit is reached. There is some evidence that a worsening of load factors in recent years (from an estimated average of 66·5 per cent in 1958 to 57·5 per cent in 1972) may be the result of the production of a greater proportion of bulky, poor-loadability goods.[2]

A final factor which may influence amalgamation possibilities is the relative use made of 'own-account' or 'hire and reward' haulage. 'Own-account' haulage can provide a high quality transport service because the vehicle's main function will be to carry the trader's own goods. This means that delivery times are likely to be reduced and losses and damage to consignments minimized.[8] The routes on which different consignments are to be carried will seldom coincide exactly, so that the amalgamation of consignments will involve some delays and diversions. For small consignments, amalgamation usually demands transhipment from a collection and delivery to a trunk vehicle. There may thus be a trade-off between 'quality of service' and achieving high load factors (and low vehicle mileage). This is a matter which can probably best be resolved by the use of the price mechanism and the retention of individual freedom in decision-making. Traders should themselves be the best judges of what it is worth while for them to pay to enjoy the high-quality, but high-cost, haulage which 'own-account' operation can supply. So long as the price which they must pay reflects the full private and social costs involved this is an area in which the market mechanism is likely to work better than state intervention. The 'quantity' licensing provisions of the 1968 Transport Act, which would have restricted the freedom to choose between 'own-account' operation and the use of the railways, would have led to great difficulties in giving a tribunal the duties of assessing the widely differing advantages of 'own-account' operation for different industries and different firms.

'Own-account' operation does not necessarily rule out all amalgamation of loads. 'Own-account' operators may welcome opportunities to carry some traffic for other traders, and particularly may be glad to reduce their transport costs by carrying back loads of traffic which is compatible with the nature of their vehicles. In a sample of 125 firms in the west Midlands, made in 1967–8, the representatives of 32 firms said that they would seek 'hire or reward' traffic when they were legally permitted to do so.[8] Since the 1968 Transport Act came into force 'own-account' operators in Britain have been free to carry goods for other traders. It is current EEC transport policy that the legal distinction between 'own-account' and 'hire or reward' operation should continue, and that 'own-account' operators should only be allowed to carry their own goods. The reimposition of this ban in Britain would make it more difficult to amalgamate loads, would be likely to lead to an increase in vehicle mileage, and would also tend to raise fuel consumption and transport costs.

## 3 Forecasts of the demand for road freight transport

The Tulpule-Tanner model used for forecasting the demand for freight transport involves making a number of assumptions. These relate to the relationship between gross domestic product and total freight ton miles, the share of the total freight market which will be carried by road, and future changes in the average carrying capacity and load factors of vehicles. If is also, of course, necessary to make assumptions about the exogenous factor, the expected rate of growth in gross domestic product. It has been shown in the previous section how past relationships and trends may change. Changes in the location of industry and of people may affect the relationship between GDP and transport demand. Since not all industries generate the same demand for transport, variations in the make-up of GDP will influence transport demand. Tanner allowed for the relatively greater growth of low transport demand industries by assuming that the ratio of ton miles to GDP will fall at the rate of 1 per cent a year.[2] A study by the Department of the Environment showed that in 1968 85–90 per cent of the demand for road goods transport came from seven industrial groups: materials for building and construction; food, drink and tobacco; solid fuel; metals; petroleum; chemicals; and machinery.[5] Changes in average carrying capacity will be affected by all those influences discussed in Section 2 in relation to the possibility of amalgamating loads. Tanner assumed that the previous trend towards higher average carrying capacities of vehicles will continue, but at a slower rate than in the past. He also made the interesting assumption that the rate of growth of carrying capacity would be positively associated with that of GDP. There is no reason why this association should exist except the belief that high GDP growth rates would lead to strong pressures to increase average loads. Hauliers might become more willing to replace smaller vehicles with larger ones and maximum permitted gross vehicle weights might be raised. But these are only suppositions and it is quite possible that there will be no association between GDP growth rates and changes in the size of average loads. Another assumption made by Tanner, which he admits is little more than a guess, is that the past decline in rail freight carrying will be reversed, and that rail freight will grow between 1972 and 2010 by 0·5 per cent per year. Tanner's forecasts of vehicle kilometres and total numbers of lorries for the years 2000 and 2010 are shown in table 3. The three levels of GDP growths used were 2 per cent, 3 per cent and 4 per cent per annum, or 'Low', 'Medium' and 'High' respectively.

**Table 3 Forecasts of vehicle kilometres and numbers of lorries in Great Britain (Tanner)**

| | Vehicle kilometres (thousand million) GDP growth | | | Numbers of lorries (millions) GDP growth | | |
|---|---|---|---|---|---|---|
| Year | Low (2%) | Medium (3%) | High (4%) | Low (2%) | Medium (3%) | High (4%) |
| 1972 | *19·83* | *19·83* | *19·83* | *0·6* | *0·6* | *0·6* |
| 2005 | 28·7 | 34·0 | 39·3 | 0·8 | 0·9 | 1·0 |
| 2010 | 30·9 | 37·5 | 44·4 | 0·8 | 1·0 | 1·1 |

Source: J. C. Tanner, *Forecasts of vehicles and traffic in Great Britain* (1974 Revision TRRC Labs. Report 650, 1974).

Although we may not be able to guess any better than Tanner it is instructive to carry out a small-scale sensitivity analysis and to examine the effect of varying his assumptions. Since Tanner has three different assumptions on GDP growth, only the effect of varying the assumptions on the growth of rail traffic, on average carrying capacity, and on load factors have been calculated. In moving from forecasts of vehicle kilometres to those of numbers of lorries Tanner made estimates of the average vehicle mileage of lorries. These estimates (which vary with GDP growth rate) have not been altered. In table 4 the total vehicle kilometres and numbers of lorries are estimated for 2010 assuming (i) that rail freight remains constant at its 1972 level; (ii) that it is reduced to zero; and (iii) that the 1972 figure is doubled by 2010. Tanner's figures are also repeated, slightly adjusted for consistency since our recalculation showed a small discrepancy, probably caused by our deduction of a constant 2 per cent to allow for light vans.

**Table 4 Forecasts of lorry kilometres and numbers of lorries in 2010 with different assumptions about rail freight**

| Rail freight assumptions | Vehicle kilometres GDP growth Low | Medium | High | Number of lorries GDP growth Low | Medium | High |
|---|---|---|---|---|---|---|
| | thousand million vehicle km | | | millions | | |
| Tanner | 30·9 | 37·3 | 44·2 | 0·84 | 0·98 | 1·11 |
| Constant 1972 | 31·8 | 38·1 | 44·9 | 0·86 | 1·00 | 1·13 |
| Zero | 35·9 | 41·7 | 48·2 | 0·97 | 1·10 | 1·21 |
| Doubled | 27·7 | 34·4 | 41·7 | 0·75 | 0·90 | 1·05 |

In table 5 the effects of changing assumptions about average vehicle carrying capacities are shown. According to Tanner's estimates average carrying capacity in 2010 would be 11·6 tonnes, 13·0 tonnes and 14·6 tonnes with low, medium and high GDP growth rates respectively. Our assumptions were that each of the carrying capacities associated with Low (11·6 tonnes), Medium (13·0 tonnes), and High (14·6 tonnes) GDP growth rates by Tanner would apply to all growth rates, and that carrying capacity would remain constant at the 1972 level (7·5 tonnes).

If the 1972 load factor of 57·5 per cent was still achieved in 2010, instead of falling to 47·5 per cent as expected by Tanner, then the 2010 ton kilometre figures (if all Tanner's other assumptions were unchanged) would be:

| | | | | | |
|---|---|---|---|---|---|
| Low | 25·6 | thousand | million | vehicle | kilometres |
| Medium | 30·8 | ,, | ,, | ,, | ,, |
| High | 36·5 | ,, | ,, | ,, | ,, |

These figures show the great importance of average carrying capacity in influencing vehicle mileage, and the relative unimportance of any

**Table 5 Forecasts of lorry kilometres and numbers of lorries in 2010 with different assumptions about average carrying capacity of vehicles**

| Carrying capacity assumed tons | Vehicle kilometres GDP growth Low | Medium | High | Number of lorries GDP growth Low | Medium | High |
|---|---|---|---|---|---|---|
| | thousand million vehicle km | | | millions | | |
| 11·6 (Low) | 30·9 | 41·8 | 55·7 | 0·84 | 1·10 | 1·40 |
| 13·0 (Medium) | 27·6 | 37·3 | 49·7 | 0·75 | 0·98 | 1·25 |
| 14·6 (High) | 25·1 | 33·2 | 44·2 | 0·68 | 0·87 | 1·11 |
| 7·5 (1972) | 47·9 | 66·0 | 87·9 | 1·30 | 1·73 | 2·21 |

likely changes in the quantity of freight carried by rail. If the 'Low' average carrying capacity was applied to the 'High' GDP forecast of road freight, vehicle kilometres in 2010 would be increased from 44·2 to 55·7 thousand million. Vehicle mileage will decrease proportionately to an increase in average carrying capacity (if Tanner's assumption that load factor is independent of carrying capacity holds). Even if rail freight carrying fell to zero in 2010, vehicle kilometres would only rise from 37·3 (on Tanner's assumption of 0·5 per cent annual growth) or 38·1 (assuming constant 1972 level of rail freight) to 41·7 thousand million, with medium GDP growth. On 1972 figures a 10 per cent increase in rail freight would reduce vehicle miles by about 2·6 per cent (assuming that all the rail increase came from the road share of total freight). A 10 per cent increase in average vehicle carrying capacity would reduce vehicle mileage by 10 per cent while, according to Tanner, a 10 per cent increase in GDP would increase vehicle mileage by 6·7 per cent.

In considering the possibility of reducing the demand for freight transport in Britain it is interesting to compare the British position with that in other countries. Although the ton mileage of freight transport per person is roughly proportional to income per head in advanced industrial countries there are some discrepancies. Sweden, Finland and the United States have a much greater ton mileage of freight per head than their income position would suggest. Table 6 shows the ratio of estimated tonne kilometres of freight to GDP in some advanced countries.

Britain is much more heavily dependent on road haulage than most other advanced countries. In 1971, 77·1 per cent of inland freight tonne kilometres in the U.K. were carried by road. The equivalent figures for West Germany and the U.S.A. were 25·6 per cent and 21·3 per cent (1969 figure) respectively.[10] Nevertheless some countries like the U.S.A. and Sweden produce more than twice as much road freight per head of the population as does Britain. No research has been carried out to explain these international variations but it would seem likely that large, less densely populated countries may have a relatively high demand for transport in relation to GDP. Some countries may also have a high proportion of those industries which generate a large demand for freight transport. It may also be significant that two of the countries with a high tonne km/GDP ratio, the U.S.A. and Sweden, also have high levels of income per head. If the relationship between income per head and ton mileage per head is not linear beyond some income level then this would mean that Tanner's assumptions of a falling ratio of ton mileage to GDP could prove wrong in some developed countries, including Britain. If there was a 'take-off' point beyond which transport per head increased at a greater rate than income per head, then the future demand for freight transport could prove to be much greater than at present expected. This is an area of great uncertainty but Britain's position as a relatively low generator of freight transport does suggest that it may prove difficult to slow down future growth in the demand for the carriage of goods in this country.

**Table 6 Ratios of tonne km to GDP, 1970**

| Country | Tonne km/GDP (in U.S. $) |
|---|---|
| Italy | 1·1 |
| Netherlands | 1·1 |
| France | 1·2 |
| Norway | 1·2 |
| G. Britain | 1·3 |
| Belgium | 1·4 |
| W. Germany | 1·6 |
| Sweden | 2·3 |
| Finland | 2·4 |
| Japan | 2·4 |
| U.S. | 3·0 |
| Australia | 4·0 |

Sources: A. H. Tulpule, *Trends in transport of freight in Great Britain* (RRL Report LR 429, 1972); J. C. Tanner, *Forecasts of vehicles and traffic in Great Britain* (1974 Revision TRRC Labs. Report 650, 1974).

## 4 The demand for passenger transport

The total demand for passenger transportation is likely to be more flexible and more easily influenced by short-term factors than is the demand for freight transport. Freight ton mileage will vary with the level of economic activity but demand will tend to be price inelastic.

Most flows of goods traffic are essential to the business of traders and cannot easily be reduced when haulage costs increase or expanded when they fall. Some passenger transport, such as the journey to work or to school, is essential, and demand will not change significantly in the short run. But some passenger travel such as pleasure rides in cars or personal social travel can quite easily be foregone or extended and may therefore be expected to be price elastic. Similarly the demand for passenger transport is also likely to be income elastic. As people grow richer their total travelling tends to increase (although travelling by some forms of public transport, which has the nature of an inferior good, may be replaced by car travel). Passenger transport demand can also be generated by the supply of transport facilities. The opening of a new motorway may persuade people to make trips to the sea or into the country which they would not have considered when the journey took much longer. There are, it may be argued, two forms of cost involved in passenger travel – money costs and time costs – and demand is likely to be elastic in relation to both of these 'prices'. The main purposes of passenger travel were analysed in a sample survey of travel made in 1972/3, and the results are shown in table 7.

**Table 7 The purpose of passenger travel, 1972/3 (all households)**

| Travel purpose | Percentage of journey stages | Percentage of mileage |
|---|---|---|
| To and from work | 29·0 | 22·0 |
| In course of work | 4·0 | 11·0 |
| To and from school and college | 7·0 | 3·0 |
| Shopping and personal business | 23·0 | 14·0 |
| Entertainment, sport, eating and drinking | 9·0 | 9·0 |
| Personal social travel | 16·0 | 18·0 |
| Other personal travel (holidays, pleasure) | 12·0 | 22·0 |
| | 100·0 | 99·0 |

Source: Department of the Environment, *National Travel Survey, 1972/73* (HMSO, 1975).

These sample figures show that 31·7 per cent of journey stages and 43·3 per cent of passenger mileage (by all inland transport modes) was for pleasure or social purposes.

As with goods transport, the measurement of passenger transport activity which is most closely associated with environmental impact is vehicle (or rail or aircraft) miles. The air pollution created by a car will not be significantly affected by the number of passengers it contains.

The noise nuisance of two half-full aircraft will be virtually double that of one plane with a full load of passengers. Table 8 shows some recent trends in passenger transport in Britain. For the purpose of comparing the market shares of different modes, passenger mileage figures must be used.

**Table 8 Trends in passenger transportation in Britain (estimated)**

| | Passenger miles (thousand million) | | | | Percentage of inland transport market | | | |
|---|---|---|---|---|---|---|---|---|
| Mode | 1952 | 1968 | 1970 | 1973 | 1952 | 1968 | 1970 | 1973 |
| Car | 37·9 | 177·7 | 196·2 | 223·5 | 33·8 | 75·3 | 77·3 | 79·9 |
| Bus and coach | 50·1 | 36·3 | 34·1 | 33·8 | 44·6 | 15·4 | 13·4 | 12·0 |
| Rail | 24·1 | 20·8 | 22·2 | 21·9 | 21·5 | 8·8 | 8·8 | 7·8 |
| Air | 0·1 | 1·2 | 1·2 | 1·5 | 0·1 | 0·5 | 0·5 | 0·5 |
| Total | 112·2 | 236·0 | 253·7 | 280·7 | 100·0 | 100·0 | 100·0 | 99·9 |

Source: Department of the Environment (formerly Ministry of Transport), *Passenger Transport in Britain* (HMSO, various years).

The demand for passenger transport may be analysed in terms of prices, incomes, and the prices of alternative goods and services[13] or by building up a predictive model for a particular mode. Tanner has developed a 'generalized' cost model to predict numbers of cars and car mileage, the costs included being those of fuel, licences, other running costs, purchase, and time.[2] Car ownership per person was estimated as a function of generalized cost and GDP (which was presumably used as a proxy for household income). Numbers of cars were then estimated from assumptions about population growth, and total car mileage was estimated on the basis of assumptions about annual car mileage. Tanner used the same three levels of GDP growth as in the freight demand study (2, 3 and 4 per cent) and also different assumptions about increases in costs. Three alternative sets of forecasts were given, 'Low', representing low GDP growth combined with high cost growth; 'Middle', combining medium cost and medium GDP growth; and 'High' which combined high GDP and low cost growth. The results of Tanner's calculations are shown in table 9.

These predictive models, and the theoretical economic analysis of the demand for passenger transport, do not tell us very much about the basic needs which lead people to want to travel. But, mainly because of

**Table 9 Forecasts of numbers of cars and car kilometres: Great Britain**

| | Number of cars (millions) | | | Annual car kilometres (thousand million) | | |
|---|---|---|---|---|---|---|
| Year | Low | Middle | High | Low | Middle | High |
| 1972 | 12·7 | 12·7 | 12·7 | 184·3 | 184·3 | 184·3 |
| 2000 | 23·1 | 24·5 | 25·1 | 353 | 385 | 405 |
| 2010 | 25·0 | 25·9 | 26·2 | 389 | 418 | 438 |

Source: J. C. Tanner, *Forecasts of vehicles and traffic in Great Britain* (1974 Revision TRRC Labs. Report 650, 1974).

the differences in demand elasticities, they are more useful as indicators of possible policy measures for reducing demand, than are the corresponding freight demand models. Even in the short term the demand for passenger transport can be reduced by raising either money prices or 'time cost' prices. In Britain the state has various means by which travel costs can be raised. Fuel costs can be raised by increasing the level of taxation. The cost of purchasing vehicles can be raised by increases in VAT, and annual running cost could be raised by increasing vehicle licence duty. Subsidies could be removed from rail and bus passenger services, and if this was not sufficient to bring about any required drop in demand, taxes similar to those applying to cars and lorries could be imposed. The impact of some of these fiscal measures would differ. In particular it is only fuel tax which varies directly with mileage and which will therefore tend to reduce annual mileage per vehicle. Increases in vehicle licence duty or VAT on vehicle purchase may result in the more intensive use of a smaller stock of vehicles. The state can increase travelling times (and thus the time cost element in travelling) by direct action such as increasing the 'red' time on traffic lights. This is only likely to happen in practice in cases like the Nottingham traffic constraint experiments where it is intended to favour one mode at the expense of another (an issue discussed briefly at the end of this Chapter). The policy which has made a strong appeal to government in Britain recently is that of lengthening, or failing to shorten, travelling times by neglecting to build or improve roads. Cut-backs in the road building programme must lead to slower journey times than those which would have resulted from a more generous road investment policy and thus cause an increase in time costs.

Since we have very little knowledge of the money value of the environmental costs imposed by passenger transport it is generally impossible to say whether a tax-induced reduction in demand will yield a net increase in welfare. Governments tend to take action to obtain calculable benefits, such as reducing Britain's bill for imported fuel, rather than to be concerned with intangible environmental goods. There will be different price elasticities for different parts of the demand for passenger transport, and it is possible to forecast which areas would be most likely to respond to a significant price increase. Pure pleasure motoring, when the only object of the journey is to look at the country-side, is clearly likely to be reduced by an increase in its marginal cost. The load factors of private cars, where the marginal cost of an extra passenger is virtually zero, are also likely to be increased when motoring costs per mile are raised. Some shift to public transport may also take place, as, even if public transport has not been excluded from the tax increase, the extra cost per passenger is likely to be lower. The long-term effects of permanently raised passenger transport prices are much less predictable. People might move nearer to their workplaces but the decision on where to live is a complex matter involving not only the cost of the work journey but also housing costs, other travel costs, and important environmental factors. It is unlikely that human happiness would be increased if large numbers of people were forced to move back from the country and from suburbs into high-density urban housing in order to reduce the environmental costs of transport. Long-term transport demand will probably be influenced by government policy on industrial and residential location and on the provision of transport services. It may also be considered desirable to discourage some travellers but not others, in a way which could not be achieved through taxation alone.

Some idea of the scope for reducing the demand for transport on the journey to work can be obtained by studying the statistics on work journeys which were produced as part of the 1966 Sample Census.[14] The figures show very considerable differences between towns in their work flow situations. Some towns 'export' large numbers of workers to other areas, while others are large-scale 'importers'. Table 10 shows the ten largest 'exporters' amongst county boroughs and GLC boroughs; the equivalent figures for 'importers' are shown in table 11. The figures in tables 12 and 13 show the other extreme, the towns with lowest 'export' and 'import' percentages. No London boroughs came into either category.

The 'export' and 'import' figures for 1966 are combined in table 14 to show the county boroughs and London boroughs which generated the largest inward and outward flows of 'journey to work' traffic. The lowest 'traffic generators' are shown in table 15. No London boroughs were included as none of them had a combined percentage of less than 100. The lowest London borough 'generator' was Croydon, for which the combined figure was 117·7 per cent.

The interpretation of the figures in tables 10–15 is not entirely straightforward. The statistics relate only to work journeys crossing local authority boundaries and it is apparent that some of the differences between towns merely reflect the accidents of these areas. Some of the former county boroughs had tightly-drawn boundaries which excluded residential suburbs, most of whose inhabitants had to cross into the borough to work. These work journeys were not necessarily any longer than those in towns with boundaries which included all the main residential suburbs. The 'import' figures for Derby were reduced from 86·1 per cent to 27·1 per cent between 1966 and 1971 because its

**Table 10 Ten largest 'exporters' of workers, county boroughs and London boroughs, 1966 and 1971**

| | Workers living in town but working elsewhere as percentage of number of workers living and working in town | |
|---|---|---|
| | 1966 % | 1971 % |
| *County boroughs* | | |
| 1. Solihull | 152·6 | 134·3 |
| 2. Wallasey | 135·5 | 112·5 |
| 3. Bootle | 98·2 | 95·7 |
| 4. Salford | 82·6 | 80·4 |
| 5. Birkenhead | 78·6 | 77·1 |
| 6. Dudley | 69·3 | 75·0 |
| 7. Gateshead | 68·0 | 61·4 |
| 8. Middlesbrough | 61·0 | —— N. a. |
| 9. Warley | 57·9 | 53·0 |
| 10. Dewsbury | 55·7 | 65·3 |
| *London boroughs* | | |
| 1. Kensington | 184·1 | 172·6 |
| 2. Lewisham | 172·7 | 168·6 |
| 3. Lambeth | 164·7 | 177·5 |
| 4. Redbridge | 164·6 | 163·9 |
| 5. Harrow | 154·2 | 150·3 |
| 6. Wandsworth | 151·7 | 152·7 |
| 7. Harringay | 147·0 | 154·0 |
| 8. Bexley | 139·4 | 145·9 |
| 9. Merton | 135·4 | 136·2 |
| 10. Hammersmith | 132·9 | 140·4 |

Source: *Sample Census 1966*, Workplace and Transport Tables, Part II (HMSO, 1968), and *1971 Census*, Workplace and Transport Tables, Part II (HMSO, 1971).

**Table 11 Ten largest 'importers' of workers, county boroughs and London boroughs, 1966 and 1971**

| | Workers travelling from outside to work in town as percentage of workers living and working in town 1966 % | 1971 % |
|---|---|---|
| *County boroughs* | | |
| 1. Bootle | 123·1 | 155·8 |
| 2. Newcastle upon Tyne | 98·5 | 118·2 |
| 3. Gateshead | 95·1 | 111·6 |
| 4. Wakefield | 93·5 | 105·9 |
| 5. Salford | 86·5 | 80·4 |
| 6. Derby | 86·1 | 27·1 |
| 7. Oxford | 79·2 | 90·5 |
| 8. Doncaster | 78·7 | 102·0 |
| 9. Manchester | 75·4 | 87·9 |
| 10. Wigan | 75·0 | 75·4 |
| *London boroughs* | | |
| 1. City of London | 13,063·5 | 14,977·9 |
| 2. Westminster | 537·4 | 561·7 |
| 3. Camden | 293·8 | 356·5 |
| 4. Islington | 194·6 | 218·9 |
| 5. Kensington | 161·2 | 167·5 |
| 6. Tower Hamlets | 140·3 | 160·7 |
| 7. Southwark | 129·2 | 142·8 |
| 8. Lambeth | 126·0 | 146·0 |
| 9. Hammersmith | 123·5 | 135·2 |
| 10. Barking | 107·2 | 120·7 |

Source: *Sample Census 1966*, Workplace and Transport Tables, Part II (HMSO 1968), and *1971 Census*, Workplace and Transport Tables, Part II (HMSO, 1971).

**Table 12 Ten towns with lowest 'export' of workers as a percentage of workers living and working in town, 1966 and 1971**

| Town | 1966 % | 1971 % |
|---|---|---|
| 1. Barrow | 5·0 | 4·2 |
| 2. Plymouth | 5·6 | 5·3 |
| 3. Oxford | 6·0 | 8·2 |
| 4. Sheffield | 6·9 | 7·3 |
| 5. Exeter | 7·4 | 7·0 |
| 6. Yarmouth | 8·1 | 7·4 |
| 7. Coventry | 8·3 | 12·7 |
| 8. Birmingham | 8·7 | 11·2 |
| 9. Hull | 9·2 | 12·5 |
| 10. Northampton | 9·3 | 10·2 |

**Table 13 Ten towns with lowest 'import' of workers as a percentage of workers living and working in town, 1966 and 1971**

| Town | 1966 % | 1971 % |
|---|---|---|
| 1. Hastings | 9·7 | 13·1 |
| 2. Southport | 13·6 | 15·9 |
| 3. Barrow | 15·7 | 21·0 |
| 4. Blackpool | 18·2 | 20·9 |
| 5. South Shields | 19·2 | 24·9 |
| 6. Hull | 19·6 | 24·2 |
| 7. Plymouth | 19·6 | 14·1 |
| 8. Northampton | 21·4 | 28·8 |
| 9. Carlisle | 21·7 | 26·3 |
| 10. Leeds | 22·6 | 29·2 |

**Table 14 Ten largest traffic generators, county boroughs and London boroughs, 1966**

| | 'Imported' and 'exported' workers as a percentage of those living and working in the town % |
|---|---|
| *County boroughs* | |
| 1. Bootle | 221·3 |
| 2. Solihull | 216·6 |
| 3. Salford | 169·1 |
| 4. Gateshead | 163·1 |
| 5. Wallasey | 160·2 |
| 6. Birkenhead | 138·0 |
| 7. Wigan | 128·8 |
| 8. Wakefield | 127·0 |
| 9. Newcastle upon Tyne | 124·1 |
| 10. Warley | 121·6 |
| *London boroughs* | |
| 1. City of London | 13,100·0 |
| 2. Westminster | 586·4 |
| 3. Camden | 412·7 |
| 4. Kensington | 345·3 |
| 5. Islington | 309·1 |
| 6. Lambeth | 290·8 |
| 7. Hammersmith | 256·4 |
| 8. Southwark | 236·3 |
| 9. Redbridge | 224·6 |
| 10. Lewisham | 222·8 |

**Table 15 Ten towns generating the smallest amount of cross-boundary 'journey to work' traffic, 1966**

| Town | % of workers living and working in town |
|---|---|
| 1. Barrow | 20·8 |
| 2. Hastings | 24·9 |
| 3. Plymouth | 25·4 |
| 4. Hull | 28·8 |
| 5. Northampton | 30·7 |
| 6. Sheffield | 30·8 |
| 7. Exeter | 31·7 |
| 8. Leeds | 32·6 |
| 9. Eastbourne | 32·7 |
| 10. Ipswich | 35·4 |

boundaries were extended to include suburbs formerly outside the county borough. Nevertheless it is highly probable that journeys involving crossing local authority boundaries would generally be longer than those made within one authority's area. These longer journeys may increase vehicle mileage, and thus be environmentally costly in two ways. A three-mile journey to work by car will obviously involve three times more vehicle mileage than a one-mile journey. But it is also probable that shorter journeys will not involve the use of motorized transport at all, but will be made by cycle or on foot. This means that the relationship between the length of work journeys and the environmental impact may not be linear. Once the maximum acceptable walking or cycling distance is exceeded then the demand for motorized transport will increase steeply. An analysis of the data for all the 78 county boroughs in England and Wales in 1966 showed, as might be expected, that people travelling across local authority boundaries were much more likely to use cars for the journey to work than those living and working within the same county borough boundary.

The average figures were:

*Living in town but working elsewhere*
Using car = 35·1%
Walking = 5·2%
*Working in town but living elsewhere*
Using car = 40·4%
Walking = 3·8%
*Living and working in town*
Using car = 18·3%
Walking = 31·2%

The 'walking' figures also included a very few people travelling by some unspecified 'other' mode of travel (specified modes being train, bus, car, goods vehicle, motorcycle and pedal cycle) and also those who had no means of travel to work, presumably people who lived and worked on the same premises. Standard errors were not calculated as all county boroughs were included, but it must be remembered that the data for each town was collected on a 10 per cent sample basis. There are also a few obvious enumerative errors. Thus six people resident in the Tyneside conurbation but working in Greater London were recorded as travelling to work on foot![14]

The Sample Census statistics indicate that the journey to work creates a much bigger demand for transport services in some parts of Britain than in others. If the objective is to minimize vehicle journeys the 'ideal' town appears to be somewhere like Barrow-in-Furness, which is not part of a conurbation, and which is relatively small. Large towns like Sheffield and Leeds, even although they generate relatively little cross-border traffic, nevertheless create a high demand for motorized transport. The coefficient of correlation between the size of the local workforce (those living and working in the town) and the percentage walking to work was calculated for all county boroughs. The result was the negative correlation of $r = -0{\cdot}51$, indicating that the larger the local workforce the less are people likely to walk to work. Towns like Solihull and Wallasey, which are largely residential suburbs for large conurbations (although both also draw in some workers from outside), generate a particularly high demand for transport services. In the four towns of Barrow, Hastings, Solihull, and Bootle, the relationships between the total number of jobs (all the jobs in each town plus those of people living in the towns but working elsewhere) and the number of motorized (including rail) transport journeys were calculated as percentages. The percentages of motorized journeys were:

| | 1966 | 1971 |
|---|---|---|
| | % | % |
| Barrow | 63·8 | 69·3 |
| Hastings | 69·0 | 76·1 |
| Bootle | 77·0 | 81·2 |
| Solihull | 83·3 | 86·9 |

Thus moving from the journey to work situation in Solihull to that in Barrow would reduce the demand for motorized transport by 23·4 per cent. This does not allow for any reduction in vehicle mileage caused by shorter length journeys. If the use of cars, rather than that of all motorized transport, is related to work journeys in the same four towns then the position of Bootle is changed considerably. The figures for the percentage of work journeys made by car were:

| | 1966 | 1971 |
|---|---|---|
| | % | % |
| Bootle | 18·4 | 30·0 |
| Barrow | 23·5 | 35·5 |
| Hastings | 27·1 | 39·3 |
| Solihull | 47·8 | 58·7 |

If a serious attempt were to be made to reduce the demand for transport for the journey to work some attention might be given to those towns which have considerable inward and outward flows of workers. Four examples are Oldham, Salford, Warley, and Ealing. The figures for inflow (people working in the towns and living elsewhere) and outflow (people living in the town but working elsewhere) were:

| | 1966 | | 1971 | |
|---|---|---|---|---|
| | Outflow | Inflow | Outflow | Inflow |
| Oldham | 16,230 | 21,890 | 15,920 | 20,370 |
| Salford | 31,240 | 32,750 | 25,050 | 36,320 |
| Warley | 31,370 | 34,410 | 27,350 | 40,320 |
| Ealing | 71,780 | 68,730 | 72,500 | 62,750 |

The outflow of workers from residential suburban towns like Solihull serves a recognizable purpose in so far as Solihull provides a superior environment in which to live compared with that found in many parts of Birmingham. But it is not so obvious why approximately equal numbers of workers should travel from and into Ealing to work. It is, of course, too simple to suppose that 62,000 people living in Ealing could change houses with 62,000 workers who travel into Ealing each day and thus greatly reduce the volume of work journeys. The incoming workers may not come from the same places as those to which the outgoing workers must travel. Tastes may differ about the desirability of Ealing as a place of residence. Nevertheless, this two-way flow of workers does present a possible area in which passenger transport demand might be reduced. In the long run increasing the price of transport (relative to incomes and other prices) may reduce the amount of 'unnecessary' work journeys. It may also be worth while to examine whether the over-provision of relatively cheap public transport may not generate an excessive demand for 'journey-to-work' transportation.

## 5 Demand for air transport

Measured by percentages, air transport is the most rapidly growing form of passenger transport in Britain. If 1970 = 100 the index numbers for 1973 were:

| | |
|---|---|
| Air – domestic | 125·0 |
| Air – international | 121·4 |
| Car | 113·9 |
| Bus | 99·1 |
| Rail | 98·6 |

(The international air figures relate to 'passengers'; all the others to passenger miles.) Recent growth in the demand for international air transport is shown by the following figures of monthly averages of 'passengers uplifted' by British airlines:

| | (*thousands*) |
|---|---|
| 1968 | 420·1 |
| 1969 | 429·9 |
| 1970 | 447·1 |
| 1971 | 447·2 |
| 1972 | 490·8 |
| 1973 | 542·7 |

It seems most likely that the demand for air passenger transport will continue to grow at a rapid rate except in times of economic depression. The growth of domestic flights might be slowed down by the successful development of British Rail's high speed and advanced passenger trains. There is some evidence that the electrification of British Rail's London to Scotland route removed some potential air traffic, at least temporarily. But this is uncertain, and the demand for international flights is almost certain to increase if there is any growth in GDP per head. As incomes increase so time is valued more highly and people tend to choose faster means of transport. Technological developments producing faster or larger (and therefore cheaper) aircraft may also influence future demand.

There seems to be little scope for reducing the demand for air transport in the near future. There is now often no other alternative for long-distance travel and the decision not to construct the Channel Tunnel means that only cross-channel ferries and the hovercraft compete with aircraft as a means of transport between Britain and the Continent.

## 6 Inter-modal diversion of traffic

In most of the discussion in this Chapter demand has been treated as though it were specific to a particular mode of transport. In fact, of course, there is considerable scope for moving goods or passengers from one mode to another. Where the total demand for a particular transport service cannot be reduced it may be possible to divert people or goods from a mode with high social costs to one which makes less impact on the environment.

The most commonly argued cases are for moving freight from road to rail and people from cars into buses or trains. The possible role of the railways in reducing the environmental costs of transport is such an important issue that it is treated separately in Chapter Nine. The case for persuading or compelling people to travel by bus rather than car has given rise to an extensive literature, which it is not intended to review here.[15,16] The main arguments for promoting travel by bus is that buses use road space more economically and thus reduce road congestion. As was explained in the introduction, consideration of the problem of road congestion has largely been excluded from this study. To what extent there are environmental benefits in having a smaller number of buses rather than a larger number of cars is a debatable issue on which little research has been carried out. A bus will normally impose heavier environmental costs than an average car. It is likely to be noisier, cause more road wear and vibration and to be more visually intrusive. But how many cars have the same environmental impact as one bus is an unresolved question. The ratio would vary for different environmental costs. It would take a large number of medium-size cars to cause the same road wear as one double-decker bus but two cars are likely to be more of an accident hazard or a barrier to crossing the road than one bus. The comparison is complicated because buses and cars are unlikely to follow the same route. If buses gave a door-to-door service they might often do more damage to the environment than cars. But a transfer from a car to a bus journey usually means substituting walking for part of the car journey. This will result in lower environmental costs but provide what most people would consider to be an inferior transport service. It is perhaps interesting to note that current transport policies in Britain tend to favour having a larger number of small lorries rather than a smaller number of large ones, but to encourage the use of a smaller number of larger passenger carriers (buses) rather than a larger number of smaller ones (cars).

**Summary**

*1. The environmental impact of transport will vary with vehicle mileage rather than with the actual volume of goods and people needing transportation.*

*2. The vehicle mileage for any given level of demand for goods transportation can be reduced by shortening the length of haul, or by increasing average loads per vehicle (or railway wagon).*

*3. Average length of haul might be reduced by dispersing manufacturing activity in some industries and by arranging marketing agreements between firms.*

*4. The average carrying capacity of vehicles has an important influence on average loads.*

*5. The removal of the right of 'own-account' operators to carry for 'hire or reward' would tend to lower average loads and increase vehicle mileage.*

*6. The extreme range of forecasts of lorry vehicle kilometres in 2010 was from 30·9 to 87·9 thousand million, but the assumptions leading to the highest figures are unlikely to hold.*

*7. The demand for passenger transport is more price elastic than that for goods transport and more likely to be influenced by pricing policy.*

*8. Forecasts of car kilometres in 2100 ranged from 389 to 438 thousand million.*

*9. There would seem to be some scope for reducing the vehicle mileage generated by work journeys in the future. Generation of vehicle mileage is already much lower in some parts of the country (particularly in smaller and more isolated towns) than in others (such as the large conurbations). Some towns, like Ealing, have very large inward and outward flows of workers and there is a possibility that people could live nearer to their place of work in these areas without suffering any loss of amenity.*

Chapter three

# Noise

## 1 Sound and noise

If beauty is in the eye of the beholder then noise is in the ear of the listener. The most useful definition of noise is that it is unwanted sound, but which sounds are wanted and which are not depends on human judgment and there is no objective test which can be used to classify sound. This Chapter is being written to the accompaniment of the roar of Cornish surf and the cry of seagulls. To most people, perhaps, neither of these sounds constitutes noise. But there are times on a rough winter night when the sound of the waves may be unwanted, and even the seagull's voice may lose its romance at six in the morning. Tastes and ears differ to such a degree that one man's music may often be another man's noise. This is true even of some transport sounds. Some sports cars are deliberately silenced incompletely because the sounds which they produce are desired by prospective purchasers, though they may be regarded as noise by almost everyone else. At a public meeting in Chelmsford, when the rival environmental benefits of road and rail transport were under discussion, one speaker said that while he hated the noise of lorries, he enjoyed hearing the trains go by. A pop group, a symphony orchestra and a jet plane may all have the same sound level in dB(A) and no noise meter so far developed can distinguish between wanted and unwanted sounds.

The reasons for disliking noise are that it may be regarded as intrinsically unpleasant; because it interferes with some wanted sound such as speech or music; because it interferes with work performance, particularly where this requires mental concentration; because it prevents sleep; and because it has physiological effects. The report of the working group on research into road traffic noise stated that 'At the present time there is no reason to suppose that exposure under normal conditions to the noise from road traffic is in any way harmful to hearing even for protracted durations of exposure nor that it has any other direct adverse physiological effect.'[1] However, evidence from some operators contacted during the *Living with the Lorry* survey suggested that lorry drivers who are exposed to cab noise for a large part of their working day may have their hearing affected.

To discuss the problem of controlling traffic noise it is necessary to understand the meaning of the relevant units of measurement. Unfortunately this is by no means easy for those who have not made a study of acoustics. The brief scientific definitions contained in some reports on transport noise are not always particularly helpful, while attempts to give a simple explanation of complex issues may be dangerous. Nevertheless it is essential for the purposes of this study to attempt some explanation of the units used for measuring transport noise. The problem of measuring the unpleasantness of noise is complicated because the reactions of human beings to noise do not depend only on the magnitude of the physical changes in air pressure. Sounds vary both in intensity and frequency and noises at some frequencies are generally felt to be more unpleasant than those at others. The threshold of hearing varies very considerably for sounds of different pitch. At 2,000 Herz (cycles per second) the ear can detect changes of air pressure as small as $10^{-3}$ micro-bar (or dyne/cm$^2$), whereas at about 24 Hz sound changes of 10·0 micro-bar can only just be heard.[2] Sources of traffic noise, such as cars and lorries, produce a complex spectrum of sound with peaks and troughs occurring at different frequencies. The basic measurement for sound intensity is the bel, or the generally used smaller unit, the decibel, written as dB. The decibel measures the ratio between the amplitude of sound pressure level and a basic reference level. This reference level, $A_0$, corresponds to the sound pressure amplitude at which sound can just be detected by the human ear at 1,000 Hz (+0·0002 dyne/cm$^2$). The decibel rating of a sound is 20 times the log of the ratio of the sound pressure amplitude to the reference level, and is given by the expressions

$$20 \log_{10} \frac{A_1}{A_0}$$

where $A_1$ is the amplitude of the sound being measured and $A_0$ is the reference level. The effect of sound on the human ear (known as the sensation level) depends upon the sound intensity (at any given frequency), which varies in proportion with the square of the amplitude. This means that the decibel measurement is related to sound intensity by the expression

$$dB + 10 \log_{10} \frac{I_1}{I_0}$$

where $I_0$ is the reference intensity (corresponding to the reference pressure amplitude of 0·0002 dyne/cm$^2$) and $I_1$ is the intensity of the measured sound.[2, 3] If all sounds were of the standard frequency of 1,000 Hz the sensation level measured in decibels would also measure the subjective impression of the relative loudness of different sounds.[4] But this is not true of noise consisting of sounds of many different frequencies, such as that produced by road vehicles. Phon and sone scales have been developed to measure loudness, but since these scales cannot at present be used in practice for measuring traffic or vehicle noise they need not be described. The problem of relating vehicle noise loudness (and presumably unpleasantness) is tackled instead by the use of the 'A' weighted decibel scale in which a weighted average decibel reading is recorded, the middle frequencies that affect the human ear most being given the largest share of the weights. Although the dB(A) scale does not vary directly with loudness in the same way as the sone scale it has the great advantage of being directly measurable with relatively simple equipment in operating or test conditions. Other weightings of the dB scale exist, and aircraft noise is measured (for no very obvious reason) on the PNdB (or perceived noise) scale, which has its own weighting pattern and cannot be compared directly with the dB(A) scale, although an approximate conversion can be made by assuming that PndB = dB(A) + 13. Subjective reaction to the dB(A) scale suggests that an increase of 10dB(A) is approximately equal to a doubling of loudness. In terms of the sound intensity ratio (and loudness for a pure tone of 1,000 Hz) an increment of 3dB represents a two-fold increase ($10 \times \log^2 = 3{\cdot}010$).

Some idea of relative *peak* noise levels from transport and other sources measured on the logarithmic dB(A) scale can be obtained from the following examples:

| | dB(A) |
|---|---|
| Jet aircraft on ground | 130 |
| Noise under flight path of supersonic aircraft within 5 miles of take off | 125 |
| Pop group | 110–25 |

| | |
|---|---|
| Noise under flight path of heavy jet within 5 miles of take off | 115 |
| Riveting machine in sheet metal shop | 115 |
| Woodworking shop | 111 |
| House near airport | 100 |
| Heavy lorry | 88–92 |
| Train | 90–92 |
| Sports car | 81–91 |
| Large car | 80–82 |
| Small car | 77–83 |
| Major road with heavy traffic | 63–75 |
| Residential roads with local traffic only | 56–65 |
| Places of few road noises and only very distant traffic noise | 47–50 |
| Quiet bedroom | 30 |
| Rustling leaves | 25–30 |
| Soundproofed broadcasting studio | 20 |

Although transport noise is undoubtedly one of the most pervasive forms of noise it seems to have received almost an undue amount of attention in Britain in relation to noise from other sources. After reading the literature on aircraft noise and road traffic noise and studying the government regulation of most forms of transport noise it is quite startling to examine the very different standards which are applied to noise at work. Traffic noise which would result in indoor noise levels of 55–60dB(A) would generally be considered to be unacceptably high. But the maximum sound level recommended (but not legally enforceable) in a government 'code of practice' for people at work is 90dB(A) for continued exposure in any one day and for a reasonably steady sound.[5] This is roughly four times louder than the traffic noise which is held to be unacceptable out of doors and eight times louder than the indoor equivalent of 60dB(A). These figures do not in any way imply that the problem of transport noise is unimportant but they do suggest that the costs of reductions of transport noise, of noise at work, and of other forms of noise nuisance should be compared.

## 2 Sources of transport noise

The main sources of transport noise in Britain are road traffic, aircraft and trains. The most universal, and in some ways the most complex, problem is that of road traffic noise. The chief variables determining the level of traffic noise are the composition of the traffic flow, the volume of traffic and traffic speed. Noise levels will also, of course, be attenuated with the distance of the road from the point of measurement (or from affected houses and people). Noise levels are also influenced by the road surface. Peak noise levels at high speeds are about 3 dB(A) higher on hot rolled asphalt or surface dressing such as that used on motorways than they would be on fine asphalt.

A road traffic noise computer model has been developed at the TRRL which includes all the variables mentioned except road surface.[6] An important assumption in this model is that it is only necessary to divide traffic into two classes. These are vehicles of up to 30 cwt unladen weight and those exceeding this weight. Petrol-driven cars and light vans fall into the former category while the heavier vehicles are mostly diesel-engined lorries and buses. For convenience the two categories are called 'cars' and 'lorries'. The justification for this simple dual classification is contained in unpublished work by Ross[7] and Lewis.[8] The computer model can be expressed in the form of simple equations which illustrate the relationships. As might be expected the proportion of lorries in the traffic flow is shown to be of greater importance in determining $L_{10}$, the 10 per cent peak noise levels, than it is in influencing the overall noise levels measured by $L_{80}$ or $L_{90}$. The basic equation for measuring $L_{10}$ (without the attenuation terms) at 10 metres is $L_{10} = \alpha(p) + \beta(p) \log_{10} V + \gamma(p) \log_{10} Q$, where p is the proportion of lorries, V is average traffic speed and Q is traffic flow. The coefficients $\alpha$, $\beta$ and $\gamma$ are not constant but are all functions of the proportion of 'lorries' in the traffic flow. But for $L_{90}$ the appropriate equation is $L_{90} = \alpha(p, Q) + \beta(Q) \log_{10} V + \gamma(p) \log_{10} Q$ and $\alpha$ is a function of both p and the traffic flow while $\beta$ is a function only of traffic flow. The output of the computer model showed that the attenuation of $L_{10}$ with distance is independent of traffic speed flow or composition, while that of $L_{10}$ is highly dependent on vehicle flow. The calibrated equation for $L_{10}$ produced by the computer model, including the attenuation terms, is

$$L_{10} = 27{\cdot}4 - 10{\cdot}3p + (20{\cdot}4 - 0{\cdot}18p) \log_{10}V + (8{\cdot}0 + 0{\cdot}05p) \log_{10} Q - 16 \log_{10}d$$

where d is the distance of the measurement point from the nearside kerb. The effect of some changes in the variables is shown in table 16.

Doubling average traffic speeds thus has considerably more effect on increasing $L_{10}$ than does doubling traffic flow. An increase in the percentage of 'lorries' has a greater effect on noise levels at lower than at higher average traffic speeds. Underlying some of these relationships is the fact that the noise created by 'cars' tends to increase with speed more than does 'lorry' noise. A somewhat different picture of the contribution

**Table 16 Effect on $L_{10}$ noise levels of increasing traffic flow, average speed, and the proportion of 'lorries'**

| Change | Increase in $L_{10}$ |
|---|---|
| | dB(A) |
| Speed (V) doubled: | |
| p = 0% | 6·1 |
| p = 40% | 4·3 |
| Traffic flow (Q) doubled: | |
| p = 0% | 2·4 |
| p = 40% | 3·0 |
| Percentage of 'lorries' increased by 40% | |
| Q = 200 veh/hr V = 100 km/hr | 2·2 |
| Q = 2,000 veh/hr V = 100 km/hr | 4·2 |
| Q = 200 veh/hr V = 50 km/hr | 4·4 |
| Q = 2,000 veh/hr V = 50 km/hr | 6·4 |

Source: P. M. Nelson, *A computer model for determining the temporal distribution of noise from road traffic* (TRRL Report LR 611, 1973).

of heavy lorry noise to peak noise levels is seen in the results of the studies of traffic conditions in three British 'High Streets', Putney, Camberley and Newbury.[9, 10] Data was collected in these surveys on the incidence of very loud individual noise peaks and on the vehicles which caused them. Some of the results are summarized in table 17.

**Table 17 'High Street' noise peaks of 85 dB(A) and the class of vehicle responsible**

| | No. of vehicles producing noise peaks of 85 dB(A) or more during sample periods | | | Total number of vehicles | | |
|---|---|---|---|---|---|---|
| Type of vehicle | Newbury | Camberley | Putney | Newbury | Camberley | Putney |
| Motorcycle | 5 | 2 | 7 | 105 | 35 | 93 |
| Car and light goods | 5 | 4 | 7 | 425 | 1932 | 2548 |
| Medium goods | 6 | 18 | 23 | 509 | 230 | 405 |
| Heavy goods | 6 | 3 | 26 | 21 | 6 | 55 |
| Buses | 2 | 3 | 20 | 138 | 79 | 184 |
| Total | 24 | 30 | 83 | 5032 | 2282 | 3285 |

Source: A. W. Christie, J. Prudhoe and M. A. Cundill, *Urban freight distribution: a study of operations in High Street Putney* (TRRL Report LR 586, 1973); A. W. Christie, R. S. Bartlett, M. A. Cundill and J. Prudhoe, *Urban freight distribution; studies of operations in shopping streets at Newbury and Camberley* (TRRL Report 603, 1973).

Table 18 gives the same data (with the same total number of vehicles) for 75 dB(A) noise peaks.

**Table 18 'High Street' noise peaks of 75 dB(A) and the class of vehicle responsible**

| Type of vehicle | No. of vehicles producing noise peaks of 75 dB(A) or more during sample periods | | |
|---|---|---|---|
| | Newbury | Camberley | Putney (80 dB(A) ) |
| Motorcycle | 45 | 21 | 14 |
| Car and light goods | 138 | 150 | 20 |
| Medium goods | 191 | 122 | 88 |
| Heavy goods | 21 | 5 | 49 |
| Buses | 102 | 58 | 95 |
| | — | — | — |
| Total | 497* | 356 | 266 |

* Total in TRRL 603 shown as 502.

Source: A. W. Christie, J. Prudhoe and M. A. Cundill, *Urban freight distribution: a study of operations in High Street Putney* (TRRL Report LR 586, 1973); A. W. Christie, R. S. Bartlett, M. A. Cundill and J. Prudhoe, *Urban freight distribution: studies of operations in shopping streets at Newbury and Camberley* (TRRL Report 603, 1973).

The data indicates, as might be expected, that medium and heavy goods vehicles are much more likely to cause individual noise peaks than are cars and light goods vehicles. In Putney, for example, 47·3 per cent of heavy goods vehicles were associated with peak noise levels of 85 dB(A) or more, whereas the equivalent figure for cars and light goods vehicles was only 0·3 per cent. But the flows of cars and light goods vehicles were so large that these classes of vehicle were nevertheless responsible for a significant number of noise peaks. Vehicles, other than lorries of 1½ tons and over GVW, were responsible for 50 per cent of the 85 dB(A) and over noise peaks in Newbury, 30 per cent in Camberley and 41 per cent in Putney. The figures for 'non-lorry' responsibility for over 75 dB(A) noise peaks were 57·3 per cent in Newbury and 64·3 per cent in Camberley. Data from the High Street studies was used to estimate the effect of changes in traffic composition on $L_{10}$. The results are shown in table 19.

Even in Putney the removal of all heavier goods vehicles would only reduce $L_{10}$ by 1·5 dB(A). Replacing heavier vehicles by double the number of smaller vehicles would have no appreciable effect on $L_{10}$ levels.

Road traffic noise will also be affected by other factors not mentioned, the most important being gradient and the frequency of intersections. Vehicles will usually be noisier when they have to engage low gear to climb hills or to start from rest. The sound level from a lorry travelling at 30 mph is increased by about 9 dB(A) on a 1 in 20 gradient. One of the reasons for the higher noise peaks of heavier vehicles is their relatively low power to weight ratios and the consequent need for frequent low-gear operation.

**Table 19 Effect on $L_{10}$ of changes in traffic composition, 'High Street' studies**

| Change in traffic composition | Change in $L_{10}$ dB(A) | | |
|---|---|---|---|
| | Newbury | Camberley | Putney |
| Removal of all goods vehicles with more than 2 axles (generally over 16 tons GVW) | −0·2 | −0·2 | −1·5 |
| Replacing goods vehicles with more than 2 axles with twice number of 10 to 16 ton vehicles | +0·1 | — | — |
| Removal of all goods vehicles of over 10 tons GVW | −0·6 | −0·8 | −2·6 |
| Removal of buses | −0·8 | −0·6 | −1·7 |

Source: A. W. Christie, R. S. Bartlett, M. A. Cundill and J. Prudhoe, *Urban freight distribution: studies of operations in shopping streets at Newbury and Camberley* (TRRL Report 603, 1973); A. W. Christie, J. Prudhoe and M. A. Cundill, *Urban freight distribution: a study of operations in High Street Putney* (TRRL Report LR 586, 2973).

Whereas traffic noise on many roads is virtually continuous during the working day the noise from aircraft and rail trains is generally that of individual planes or locomotives. As has been shown, peak noises from trains are as high as those from heavy lorries while aircraft noise can be considerably louder. Since train and aircraft noise does not merge into a general traffic noise the severity of the noise nuisance is much more dependent on the number of trains or aircraft in operation. A few more cars and lorries on a busy road will have little effect on noise levels whereas a few more aircraft operating on a particular flight path will probably result in a proportional increase in noise nuisance. Given the same percentage fall in road, rail, or air traffic it seems likely that road traffic noise levels would be reduced less than those for the other modes. This does not mean that reductions in air or rail traffic would bring greater benefits to the communities than equivalent reductions in road traffic since more people may be exposed to road noise.

## 3 Possible methods for reducing transport noise nuisance

There are a number of possible remedies for the transport noise nuisance. Some of these apply to all types of transport noise. The list of potential methods for reducing the impact of transport noise is: reduction in vehicle speeds; reduction in traffic flows (or in train or aircraft frequency); changes in traffic composition; changes in vehicle (or locomotive or aircraft) technology; changes in road construction technology; the use of sound barriers of different forms; and the separation of transport noise from people by distance. The first five 'remedies' attempt to deal with noise creation at source while the last two are methods of lessening its impact.

As shown in table 16 the main scope for significant gains from reductions in vehicle speed is on roads with flows of fast-moving cars. These roads are, however, not likely to run very close to built-up areas so that the greatest possibilities for noise reductions through reducing traffic speeds may be in areas where the demand for increased quietness is not particularly strong. It seems unlikely at present that speed restrictions will be imposed solely in order to reduce traffic noise as losses through increased journey times would probably be greater than the gains from noise reduction. But the effect on traffic noise is clearly a factor which should be taken into consideration when speed limits are being determined. It is symptomatic of the current piecemeal approach to environmental and ecological problems that lower speed limits were introduced in Britain in 1974 in order to reduce fuel consumption and improve our balance of payments position. The gains to the community through reduced accident rates and through some possible reduction in traffic noise levels were not relevant to the decision, but if the original decision was justified, may be regarded as bonuses for the community. It is not clear whether there would be any significant noise advantages in reducing aircraft or train speeds although using slower existing aircraft is likely to be less noisy than flying supersonic planes.

Reducing the number of vehicles, planes or trains will reduce transport noise just as it will reduce all other disbenefits of transport operation. The problems of reducing the demand for transport carriers have already been discussed in Chapter Two. As was shown above in the discussion of the sources of transport noise, the relationship between noise and the level of transport activity is more complex for road traffic than for trains or aircraft. According to the 'High Street' data, doubling traffic flow would only increase $L_{10}$ by 2·4 — 3·0 dB(A). The main policy conclusion suggested by the flow/noise relationship is that traffic noise nuisance can be minimized by concentrating traffic flows into major routes rather than spreading traffic over a network of urban roads. In rural areas traffic noise will have less impact if one road is used for all through traffic instead of dividing the flow between two or three different roads (assuming that all the alternative routes are equally built up). A similar relationship exists with aircraft noise where less disbenefits will result if a small number of people have many aircraft flying overhead rather than more people having relatively few planes routed over their homes.[11]

The most obvious way in which traffic composition might be changed would be to reduce the proportion of lorries and buses. The degree of benefits which would result from removing lorries is dependent on which noise nuisance measurement is considered to be relevant. According to the 'High Street' studies, removing all vehicles of over 10 tons gross vehicle weight would reduce $L_{10}$ by over 0·6 to 2·6 dB(A), while the effect of removing those of about 16 tons gross vehicle weight and over would be to reduce $L_{10}$ by 0·2–1·5 dB(A). The $L_{10}$ reduction in the 'High Street' conditions if all buses were removed would be 0·8–1·7 dB(A). The computer model showed that the effect of changing the percentage of all vehicles of over 1½ tons average weight on $L_{10}$ is dependent on the size of the traffic flow and average vehicle speed. The greatest noise reduction benefits from reducing the proportion of 'over 1½ tons' vehicles would be with high vehicle flows and relatively low average speeds. On the other hand the 'High Street' studies data suggested that half the individual noise peaks of 85 dB(A) or more were caused by medium and heavy goods vehicles. If it is true that the total demand for goods transport is not likely to be reduced and that the road share of freight transport will at least remain constant, then it will not be easy to reduce the proportion of lorries in the average traffic flow. The best that can be done is to remove lorries from built-up roads to those where noise is not a nuisance because there is no-one to hear it. The possibilities of doing this are discussed in Chapter Eight. A different issue, which will arise in contexts other than that of noise, is the choice between having a smaller number of heavier lorries or a larger number of lighter, smaller capacity vehicles. It is unlikely that $L_{10}$ can be reduced by substituting more smaller vehicles for fewer heavier ones. According to the 'High Street' studies replacing goods vehicles with more than 2 axles by twice the number of 10- to 16-ton vehicles would have no significant effect on $L_{10}$ levels in High Street conditions. The computer noise model, however, does not appear to be consistent with this conclusion since all vehicles of over 1½ tons unladen weight are treated as being identical. This means that having double the number of vehicles would increase $L_{10}$. The use of more smaller vehicles could, however, remove some of the very loud noise peaks found in the 'High Street' surveys.

Probably more promising than any of the remedies for transport noise discussed so far is the possibility of developing improved technologies for vehicles, rail locomotives, and aircraft so that they produce less noise. The production of noise by road vehicles is a complex matter, about which a great deal has been written. Sources of vehicle noise are generally divided into two types, those which vary with road speed, and those varying with engine speed (although, of course, engine speed and road speed are themselves related variables). The noise sources varying with road speed are aerodynamic noise, transmission noise, body vibration and tyre noise.[12, 13] These noise sources are referred to collectively as 'rolling noise'. Engine-speed controlled noise sources are cooling fan noise, exhaust noise, air intake noise, transmission noise, body vibration, engine mechanical noise, and engine combustion noise. If I is the noise source intensity (or rate of increase of noise), V is road speed, N is engine speed, W the width of the tyre section and B is the cylinder bore, the following relationships have been shown to exist:[12]

| | |
|---|---|
| Aerodynamic noise | $I \alpha V^{5.5}$ |
| Transmission noise | $I \alpha V^{2}$ |
| Body vibration | $I \alpha V^{0}$ |
| Tyre noise | $I \alpha V^{3} W^{3}$ |
| Cooling fan noise | $I \alpha N^{5.5}$ |
| Exhaust noise | $I \alpha N^{4.5}$ |
| Air intake noise | $I \alpha N^{4.5}$ |
| Transmission noise | $I \alpha N^{2}$ |
| Body vibration | $I \alpha N^{0}$ |
| Engine mechanical noise | $I \alpha N^{3}$ |
| Engine combustion noise: | |
| normally aspirated diesel | $I \alpha N^{3} B^{5}$ |
| Turbo-charged diesel | $I \alpha N^{4} B^{5}$ |
| Petrol engine | $I \alpha N^{5} B^{5}$ |

These relationships mean, for example, that aerodynamic noise increases by 55 decibels when road speed increases by ten times or that engine mechanical noise increases by 30 dB for a tenfold increase in engine speed. There is, in each case, a linear relationship between increase of noise and the logarithm of speed. Combustion noise from a turbo-charged diesel engine (and from a petrol engine) increases more rapidly with engine speed than that from a naturally aspirated diesel because of its smoother cylinder pressure development.[12] But it must be remembered that this same feature of smooth cylinder pressure development results in a reduction in noise levels, at lower engine speeds.

In the present state of vehicle technology it is the engine-speed controlled noise sources which are most important. A lorry in 1st gear with a vehicle speed of 7–8 mph will have the same noise level (about

83–9 dB(A) ) as the same vehicle travelling in 4th gear at 40–50 mph[12] (this presumably assumes that the vehicle is travelling on a normal road surface and does not have a loose load). Without the use of silencers the exhaust system would be the main cause of vehicle noise. With existing silencers, it is the mechanical and combustion noise of engines which is the main noise source, and most of the recent work on reducing the noise from lorries has centred around the engine. The Transport and Road Research Laboratory of the Department of the Environment has recently sponsored a five-year research project for the production of a quiet heavy lorry, which is being carried out by the TRRL itself; the Vehicle Engineering Division of the Department of the Environment; the Institute of Sound and Vibration Research at Southampton University; The Motor Industry Research Association; British Leyland; Rolls Royce; and Foden. According to a recent report of a working group on Traffic Noise of the Noise Advisory Council this research, if successful, should enable goods vehicles with a maximum noise level of 80 dB(A) to be mass produced by 1980.[14]

There are two main methods which may be used to lower the noise from diesel engines. These are enclosing the engine to reduce noise radiation, and redesigning the engine itself so that it creates less noise. Engine enclosure or encapsulation offers the great advantage that it can be applied to existing engines and vehicles. It has the disadvantages that access for servicing may be restricted, that the problems of cooling become more difficult, and that the weight of the cladding reduces possible payloads. It is uncertain exactly how much sound reduction could be obtained from encapsulation in operating conditions. Experiments carried out by Aspinell and West at MIRA showed that a noise level of 80 dB(A) could be 'approached' with extensive engine and gearbox enclosure while 87 dB(A) could be achieved by engine enclosure or the use of a skirted undertray.[15] These lower noise levels were only reached in the Aspinell and West experiments (in which current models of lorries were used) by adding additional exhaust silencing. These tests did not solve the problem of engine cooling nor the possible difficulty that the enclosures would deteriorate with use and would themselves require extensive maintenance. Truck and engine manufacturers interviewed during the *Living with the Lorry* study were generally not hopeful of reaching noise levels below about 86 dB(A) through engine encapsulation. The positioning of exhaust pipes and hoses on many existing engines (in 1966) made the fitting of enclosures very difficult.

The work on an engine for the 'quiet lorry' at Southampton is based on a complete redesigning of existing engines. Recent trends in engine design have resulted in noisier, rather than quieter, engines. This is because engines have been designed to be light and to take up the smallest possible amount of space. These compact engines have large bore sizes and are designed to work at high operating speeds. Both of these factors, as shown in the relationships listed above, tend to increase engine noise. The quiet engine thus involves the choice of smaller cylinder bore sizes and operation at lower engine speeds. An important factor is that noise, although increasing with cylinder bore, does not increase significantly with the number of cylinders. Additional power can thus be obtained from using extra cylinders, though, of course, weight and cost penalties may be involved. According to Professor Priede, 'engine noise is generally independent of the volume of work done per unit-time, or horsepower, but the main criterion is the operational speed'.[16] Noise can also be reduced by the use of turbochargers. Dr N. Lalor has suggested that turbocharging can reduce engine noise, giving more power for the same pressure, and argues that all heavy lorries will have turbochargers by the 1980s.[17] Professor Priede has also argued that naturally aspirated engines would be quite incapable of meeting even the existing legal noise level requirements (89 dB(A) or 102 dB(A) on the test bed) for, for example, 44-ton vehicles required to operate at 8 BHP per ton, but that turbocharged engines can provide high power at acceptable noise levels. The desirability of turbocharging is not, however, universally accepted either by engine or vehicle manufacturers or by operators.

The quiet engine being developed at the ISVR at Southampton would have a stiff load-carrying frame replacing the present 'open' bottom of the engine which would greatly reduce vibration. The quiet engine would also have built-in noise insulation, use materials which are low noise creators (such as copper manganese alloy) for gearwheels, use low-speed, small-bore long-stroke engines, and obtain increased power from the use of more cylinders. A combination of some of these measures will, according to Professor Priede and Dr Lalor,[17] produce a diesel engine rated at about 80 dB(A).

The results of some recent TRRL research based on the noise model suggest that an important contribution to reducing traffic noise can be made by reducing the noise from cars as well as that from lorries and buses.[18] The main results are summarized in table 20.

For all traffic flows, significantly greater noise reductions would be obtained if both car and lorry noise levels were reduced. It is also interesting to note that according to these calculations greater noise reduction benefits would be gained from reducing car noise rather than lorry noise, even with 20 per cent lorries, for traffic flows of up to 400 vehicles per hour. The actual percentages of medium and heavy goods vehicles in the High Street studies (measured between 08·00 and 17·30 hours) ranged from 9·9 per cent in Camberley to 14·0 per cent in Putney. The percentage of goods vehicles (other than light vans) at 22 sample census points on motorways in 1972 ranged from 11·09 (M4 east of Holyport) to 33·14 (M50 west of A38). The unweighted average percentage at the 22 sites was 22·11. The equivalent figure for all roads in Great Britain was 8·56 per cent.[19]

**Table 20 Reductions in $L_{10}$ when lorry, car and lorry, and car noise emission levels are reduced**

| Flow vehicles/hour | 5 per cent lorries | | | 10 per cent lorries | | | 20 per cent lorries | | |
|---|---|---|---|---|---|---|---|---|---|
| | 1 | 2 | 3 | 1 | 2 | 3 | 1 | 2 | 3 |
| 200 | 0·7 | 4·5 | 5·4 | 1·4 | 4·0 | 5·6 | 2·5 | 3·5 | 6·3 |
| 400 | 0·7 | 4·5 | 5·4 | 1·4 | 4·0 | 5·6 | 2·5 | 3·1 | 6·4 |
| 600 | 0·7 | 4·5 | 5·4 | 1·5 | 4·0 | 5·8 | 2·9 | 2·1 | 6·7 |
| 800 | 0·7 | 4·4 | 5·4 | 1·5 | 3·7 | 5·9 | 3·1 | 1·1 | 7·0 |
| 1,000 | 0·7 | 4·3 | 5·5 | 1·6 | 3·2 | 6·1 | 3·5 | 0·7 | 7·4 |
| 2,000 | 1·1 | 3·5 | 5·8 | 2·4 | 1·2 | 6·8 | 4·2 | 0·5 | 7·8 |

1 = All lorries reduced by 10 dB(A)
2 = All cars reduced by 5 dB(A)
3 = All lorries reduced by 10 dB(A) and all cars by 5 dB(A)

Even if substantially quieter vehicles are manufactured it would take some years for the existing stock of vehicles to be replaced. According to the TRRL research on the introduction of quieter vehicles, $L_{10}$ levels will not be reduced by more than 2 dB(A) until at least 50 per cent of the stock of lorries have had their noise output lowered by 10 dB(A).[18]

As with lorries or cars, aircraft can be quietened either by trying to reduce the noise from existing engines or by using a basically quieter engine design. Planes in the 'second generation' of subsonic jets are substantially quieter than the earlier jet aircraft. The most important improvement has been the use of a secondary jet of cold air which reduces the noise caused by the injection of hot gases from the main propulsive jet, without causing any loss of power. Aircraft such as the Boeing 747, the Tri Star and the DC10 use these greatly improved lower noise engines. The 100 PNdB (equivalent to 87–90 dB(A)) noise contour for the 'first generation' DC8 on contact with the runway extends for about 17km (horizontally) on take-off and about 12 km on landing, while for a DC10 the equivalent figures are 6 km and 3 km. The use of supersonic aircraft like Concorde probably means moving back to higher noise levels. In order to achieve faster speeds it will be necessary to accept a higher level of noise emission.

As has been shown, peak noise levels from road traffic can be reduced by changes in methods of road construction. Reductions of about 3 dB(A) can be obtained by using road surfaces such as fine asphalt which minimize tyre noise. But the choice of road surface must also be influenced by the cost and availability of raw materials and by the expected road life given by different surfaces. Road surfaces which are low noise producers when they are in good repair may become noisier with wear. Vibration (which is discussed in Chapter Seven) will also be increased on road surfaces which wear and cause irregularities. As with all the other possible remedies noise minimization is only one objective to be considered, and it may conflict with other environmental aims. Noise creation, especially by low-powered heavy lorries and buses, will also be reduced if roads can be engineered to produce easy gradients (the noise produced by a heavy lorry travelling at 30 mph will increase by about 9 dB(A) on a 1 in 20 gradient).

Insulation against road traffic noise can be provided by sound barriers. A barrier 30 metres high, built at the edge of a 30-metre carriageway, would, it has been estimated, reduce sound levels by 12 dB(A) at 6 metres and by 8·5 dB(A) at 65 metres from the barrier. But sound barriers may be unsightly both for local residents and for those using the road. They could prove an accident hazard and, of course, on many roads in built-up areas, there is no room for their construction.

A more generally applicable form of sound insulation is the sound-proofing of houses. This will be effective against both road traffic and aircraft noise – and will also exclude or reduce some non-transport noises which disturb people in their homes such as the radio sets of neighbours or the sound of children at play. The most commonly used method of sound insulation for houses is the double glazing of windows. Double glazing can reduce the noise levels inside a house by about 10–14 dB(A). It also yields the important additional 'bonus' of reducing heat loss and thus cutting fuel bills. But a major disadvantage of double glazing is that most of the benefits will disappear when windows are opened, and not many people in Britain will want to pay for a full system of air-conditioning.

The last of the 'remedies' for transport noise is to separate noise sources from people. The creation of noise is not a problem if there is no-one to hear it. As the TRRL noise model shows, there is a regular attenuation of noise with distance, though the rate of attenuation differs for $L_{10}$ and $L_{90}$. $L_{10}$ falls by about 4·7 dB(A) when the distance from the road is doubled, while the noise from individual vehicles falls by about 6 dB(A).

An obvious advantage of building roads that will take traffic away from people is that noise from all forms of traffic, not only that of heavy lorries or buses, will be removed. The impact of other forms of transport disbenefit such as vibration, air pollution and visual intrusion, will also be reduced at the same time. Most of the other remedies discussed relate only to the single disbenefit of traffic noise. But it may be very difficult, in Britain, to find routes for roads (or railways) which will run outside built-up areas, and now road building may itself involve unacceptable environmental costs. These issues are discussed in more detail in Chapter Eight.

A fundamental issue which was examined in the Roskill Commission study on the problem of siting a third London airport[21] was the choice between a relatively remote site, where few people would be affected by the noise under the flight paths, and a site which would be convenient for access to London. Much of the criticism of the Roskill recommendation of Cublington as the preferred site centred around the fact that, according to the Commission's figures, the value of the saving in travelling time for journeys between London and Cublington compared with that from London to Foulness was greater than the extra cost of the noise nuisance at Cublington. But the noise differences between the sites was not very great, especially when allowance was made for the possible closure of Luton if the airport was sited at Cublington, and the Commissions research teams did their best to tackle the very difficult problem of putting money values on environmental disbenefits and time savings.

Since airports must serve heavily populated areas and since aircraft noise tends to disperse over a much wider area than road or railway traffic noise, the problem of airport siting is particularly intractable.

## 4 Legal controls on noise emission

The legislation relating to the maximum noise which lorries are allowed to create in Britain is complicated and confusing and is sometimes presented in an over-simplified form because qualifications and exceptions are ignored. The complications arise because different noise levels have been suggested or proposed by the Department of the Environment, by the EEC and by the ECE; because there are separate regulations relating to vehicles in test conditions and vehicles in use; and because proposed regulations have several times been modified or postponed. The current situation is that lorries now being made (and all vehicles used since 1 April 1970) should be constructed to produce a maximum dB(A) level of 89 *in ISO test conditions*. However, this regulation was modified so that new vehicles of over 200 BHP did not have to meet the 89 dB(A) level until 1 October 1974. Until that date the maximum permitted noise level was 92 dB(A). But for 'heavy' vehicles in use (over 3·5 tons gross weight) and tested on the roadside, the limit – whatever their age or brake horse power – is 92 dB(A).[22] Regulation 23 of the 1969 Construction and Use regulations, which is often quoted to suggest that all goods vehicles must conform to an 89 dB(A) maximum noise level, thus applies only to newly built lorries with engines of less than 200 BHP tested in ISO, not in operating conditions. (It should be remembered, however, that only about 1 per cent of vehicles of more than 3·5 tons weight at present have engines of more than 200 BHP;[22] similarly, the use of the word 'heavy' for all vehicles of over 3·5 tons GVW is possibly misleading since there is a very big difference between vehicles at the lower and upper end of the weight range 3·5–32 tons.) It is not clear why vehicles which can meet the 89 dB(A) level in ISO test conditions need a 3 dB(A) tolerance to allow for measurement in operating conditions whereas the over 200 BHP vehicles which cannot reach 89 dB(A) do not need, or are not given, this tolerance.

It is difficult to discover clearly the shape of future legislation. In October 1971 the Minister of Transport Industries announced new maximum noise levels, to come into operation on 1 April 1974. These levels, applying to ISO testing conditions (as defined in the British BS 3425 standard), were: light vans = 82 dB(A); 'heavy' lorries (over 3·5 tons GVW) up to 200 BHP = 86 dB(A); heavy lorries over 200 BHP = 89 dB(A). However, the implementation of these regulations was subsequently postponed for one year, and in June 1972 the Minister announced that their implementation would be postponed again and no new implementation date was given. The Minister's statement that 'these levels would have been stricter than those currently enforced within the E.E.C.' suggests a possible modification of the standards. Since these standards can be met by some vehicles already on the roads it would appear that some British manufacturers have been rather slow in developing quieter engines. The EEC limits laid down in a technical directive published in February 1970 were: cars = 83 dB(A); light goods vehicles = 85 dB(A); heavy vehicles up to 200 BHP = 90 dB(A); heavy vehicles over 200 BHP = 92 dB(A).[14] The reason given for the latest postponement of the '1974' limits was the need to coordinate with EEC proposals.

France adopted the EEC standards for cars and heavy lorries in 1972, as well as maxima for motorcycles (and mopeds) of different size. These were:

| | |
|---|---|
| Less than 50 cc | 74 dB(A) |
| 50–125 cc | 81 dB(A) |
| Over 125 cc | 85 dB(A) |

As Alexandre and Barde remark, it is strange that a motorcycle of 200 cc is allowed to make more noise than a car of 2,000 cc.

The enforcement of noise regulations in operating conditions involves considerable difficulties. Roadside conditions vary widely (from open country to noise-reflecting closely built-up areas) and there are at present no accurate portable noise meters suitable for use by the police. The great majority of successful 'noise' prosecutions (13,327 out of 14,087 in 1970) relate to noise caused by faulty silencers. There were only two successful prosecutions of vehicles for exceeding the maximum permitted sound levels in 1970, and the total fines imposed amounted to £2·00!

Aircraft constructed since 1969 must conform to standards laid down under the Aircraft Noise Certification Scheme and maximum noise levels are laid down for planes using specific airports. The noise limits for individual planes using Gatwick, Manchester and Heathrow (measured from monitoring points on the ground within a mile or two of the airport) are 110 PNdB during the day (about 98 dB(A)) and 102 PNdB at night (about 89 dB(A)). Airport noise regulations vary considerably around the world. Sydney airport, for example, bans night flights altogether. Recent research by the Consumers' Association suggests that many aircraft break the noise regulations. In sample measurements spread over several days it was found that over 70 per cent of aircraft flying over the test sites recorded noise of more than 110 PNdB.[23] The figures at Manchester and Heathrow were 79 per cent and 73 per cent respectively. This research also showed that the average NNI ratings (on sites situated in the first major housing areas under flight paths) were 50 at Luton, 60 at Manchester and 75 at Heathrow. The government-sponsored Noise Advisory Committee has suggested that NNI readings of over 35 cause annoyance and that those of 50 to 60 are intolerable. It is officially recommended that no more houses should be built in areas where the NNI reading is over 60.[23] This apparent acceptance of 'intolerable' noise levels around Heathrow and Manchester when compared with the furore created by the proposal to build the third London airport at Cublington again illustrates the somewhat haphazard nature of environmental protest and of official actions. The Consumers' Association research also examined the performance of individual aircraft and it is interesting to note that the average noise for the DC10 was 110·6 PNdB as compared with 112·4 for the DC8. This is a much less dramatic improvement for this plane than that suggested by the noise contours described earlier in this Chapter. The noisiest plane was the BAC 1-11 and the quietest (of the larger subsonic jets) was the Tri Star.

## 5 Costs and benefits

Measuring the costs and benefits of possible methods of reducing transport noise is just as difficult as determining the location of a third London airport. The possible alternatives are numerous and our knowledge of the money value of 'quietness' is very imperfect and fragmentary.

During the *Living with the Lorry* survey an attempt was made to obtain some estimates of the cost of reducing the noise created by lorries. A TRRL review quoted figures of £50–£200 for acoustic enclosures on heavy vehicle engines which would reduce noise by about 5 dB(A).[1] The average figure given by engine and vehicle manufacturers to the *Living with the Lorry* research teams for reaching 89 dB(A) on the heaviest vehicles was £185 per vehicle (at 1971/2 prices). Few firms were willing to estimate the cost of reaching 86 dB(A) for the heaviest vehicles though one figure suggested was about £1,000 per vehicle. Nobody could guess what 80 dB(A) would cost though typical comments were that it would be a 'multi-million pound exercise' or that it would involve 'a huge rise in costs'. Global costs for reducing the noise from all lorries in Britain are even more difficult to estimate. The TRRL review contained the observation that the cost of operating and providing commercial vehicles over 1·5 tons unladen weight was about £1,000 million and the statement that 'How much reduction of noise from such vehicles would cost is not known, but it is unlikely to exceed 5 per cent, i.e. £50 M.'[1] By 'reduction of noise' it would appear from the context that a fall from 89 to 86 dB(A) was meant. In the *Living with the Lorry* survey estimates of the cost of achieving maximum noise levels of 89 dB(A) for 'heavy' lorries of more than 200 BHP per ton and 86 dB(A) for all 'heavy' lorries (of 3·5 tons gross vehicle weight or more) were obtained. The figures related to new vehicles only and the average annual cost, at 1971/2 prices, was £11·9 million. The assumptions made are explained in the full *Living with the Lorry* report.[24]

The TRRL traffic noise review contained estimates of the costs of some alternative methods of reducing traffic noise. Noise barriers giving noise reductions of 5–10 dB(A) were estimated to cost £10 per foot run. The costs of constructing a motorway-standard road (dual 3-lane roads) in a cutting which would reduce the noise to nearby buildings (if they were not too high) would be about 1·3 to 4·0 times the cost of construction at ground level. (These costs do not include those of land purchase.) An estimate made in 1968 suggested that the construction costs of a motorway would increase from £0·1 million per lane mile at ground level to £5·4 million per lane mile for a motorway in a bored tunnel, though recent developments suggest rather lower costs for a tunnel.[1] The cost of screening by space depends, of course, on the cost of land which might range from £15,000 per mile in open country near London to £400,000 per 10ft width per mile in Central London.[1] The cost of sound-proofing housing by double glazing, roof insulation and door sealing was estimated to be in the range £100–£300 per house. The costs of some of these alternative noise reducing measures can be expressed in relation to the cost of reducing vehicle noise emission. Thus the cost of the '1974' noise levels of 90 and 86 dB(A) for heavy lorries would buy instead about 113 miles of roadside noise barriers (on both sides of the road); it would pay for 60 miles of motorway to be constructed in a sound-reducing cutting; or it would provide sound-proofing (double glazing, roof insulation and door sealing) for about 59,500 houses.

To measure the value of the benefits of transport noise reduction is even more difficult than that of estimating costs. As has been shown, benefits are highly subjective, and noise levels which may cause great distress to one man may be unnoticed by another. The valuation of noise disbenefits can also change quite rapidly. Some noises may become so familiar that people are no longer troubled by them, and their nuisance-value may diminish or disappear. On the other hand, as people move from noisy cities to quieter suburbs and into the country they may become less tolerant of transport noise. Another aspect of the problem is that some noises may mask other noises, so that removing the louder noise may reveal the lower noise. This revealed noise could prove just as disagreeable as the noise which was removed, so that the noise reduction benefits would be valueless. Thus, some people have objected to the whine of the transmission on buses which have been fitted with 'quiet' engines. Added to these difficulties are all the problems, discussed in Chapter One, of measuring people's evaluation of benefits which cannot be sold on the market.

An interesting British study has attempted to measure what people have paid in order to enjoy lower traffic noise levels. It was based on a survey of house prices over a ten-year period on a private estate of virtually identical houses.[25] Some of these houses were on Queslett Road, a main 'ring road' in north Birmingham (which then carried traffic from the West Bromwich-Black Country area to east Birmingham and to the east and south-east exits from that city) and others on much quieter local estate roads. At the time of the study the $L_{10}$ traffic noise levels in Queslett Road were in the range 70–5 dB(A) with 72·5 dB(A) as the most usual value at house façades. In the estate roads away from the main road the typical noise level was 52·5 dB(A). A multiple regression equation was fitted to measure the effect on house prices of a number of explanatory variables. Only four of the variables were shown to be significant. These were possession of a garage; whether the house was sold through an agent; time on the market; and noise level (measured in various ways). As was expected, the signs of the coefficients of the first and second of these variables were positive, and that of the third was negative. House prices tended to increase if houses had a garage or were sold through an agent, and to decrease the longer that the houses were on the market. What was surprising, however, was that the sign of the noise-variable coefficient was also positive. House prices tended to be higher on the main Queslett Road with its high traffic noise levels than on the quieter estate roads. Noise was the second most important variable affecting house prices (after the time the house was on the market), $L_{10}$ explaining 5·25 per cent of the variance in house prices. There were some difficulties involved in this study; house prices had to be adjusted for changes over time and data was available only for current noise levels. But the general conclusion that people were willing to pay more for houses on the main road appears to hold. As Mrs Diffey, who undertook this survey, pointed out, this does not mean that people actually enjoyed traffic noise. The most likely explanation of the association of traffic noise with higher prices is that there were other aspects of living on the main road, such as being right on the bus route and seeing more 'life', which people valued. Perhaps some families even enjoyed spotting heavy lorries! Whatever the interpretation, however, it is evident that any value which people in Kingstanding (the area of Birmingham surveyed) placed on noise reduction was not reflected in an analysis of house price differences. It would be surprising if the benefits of living on Queslett Road were sufficiently great to offset high values for 'quietness'. An American survey of the rents paid for similar apartments with different noise conditions came to the conclusion (on the basis of the statistical evidence) that although traffic noise might be a disutility, 'the analysis of the study strongly suggests that the occupant's annoyance is not reflected in rent'.[26] A very limited amount of fresh evidence on the evaluation of traffic noise reduction benefits was obtained in the *Living with the Lorry* study. Information obtained from two estate agents in Leicester about house prices was difficult to use because of the abnormal demand for houses at the time of the study. Houses were selling extremely readily, irrespective of their location. One agent did advise a client to reduce a £12,000 house in a particularly noisy situation on the Leicester ring road by £500 but this subsequently sold for £12,000. In normal times, according to both agents, more expensive houses might take longer to sell if they were on main roads. One agent said that the price obtained for relatively low-priced houses on

Narborough Road (the main A46 route from Leicester to Coventry, and link from south Leicester to the M1) was identical with that received from houses in quiet side streets in the same area. Both agents agreed that people were at least as much concerned about the problems of crossing main roads, about the safety of their children, and about the possibility of road-widening schemes taking place, as they were about traffic noise. One agent said that many people had become accustomed to noise, and that neither prices nor rents were lower on noisy roads.

A small-scale investigation was also carried out into the opinions of people living on the A426 in Blaby, on a stretch where it carried traffic between Leicester and Rugby, and between parts of Leicester and the south M1 access point. This length of road has now been by-passed. Thirty houses were visited and answers obtained to questions on traffic conditions. Seventeen respondents thought that the road traffic flow was heavy and nine that it was average. When asked what annual increase in rate payments they would be prepared to make in order to have 80 per cent of the through traffic flow removed 17 said 'none' and 12 said that they would be prepared to pay between £2 and £10 per year. One respondent had spent £100 on double-glazing so would not now valuntarily pay any rate increase. Comments from those who would pay nothing to have traffic removed included 'I like to watch the traffic'; 'lorries and traffic do not affect me'; 'I would rather have traffic than pay anything'; 'removing traffic would remove trade from Blaby'; and 'I used to live under the main flight path of London airport'. Comments from 'payers' included 'The value of the house would rise without traffic'; 'I am a night worker and traffic noise and vibration prevents me from sleeping properly'; and 'I would pay for the sake of the children'. Some of these comments relate to the total traffic nuisance rather than to noise alone.

As part of the research work for the Commission on the Third London Airport (Roskill Commission), surveys were carried out on the effect of aircraft noise on house prices.[21] The noise measurement used in the Roskill Report was the Noise and Number Index (NNI) in which aircraft noise (measured in PNdB) is related to the numbers of aircraft using particular flight paths. The NNI values are obtained from the equations:

$$\text{NNI} = \overline{\text{PNdB}} + 15 \log \text{N} - 80$$

where N is the number of aircraft heard on an average summer day (between 7·00 and 19·00 hours) and $\overline{\text{PNdB}}$ is the average *peak* noise of these aircraft. One survey compared prices obtained for individual houses at different distances from Heathrow Airport, and another was based on the 'matched sample' technique in which house values on similar estates were compared. The results of the first of these surveys showed a 6 per cent depreciation in house prices for a NNI value of 50 while the second suggested that aircraft noise at this level had no effect at all on house prices. The Roskill research team subsequently concluded that the 'matched' estates were not in fact identical, and that the higher noise level on one estate was offset by the possession of other advantages. Neither of these surveys was used in the final Roskill cost benefit calculations but a third survey was carried out based, not on actual house prices, but on the opinions of estate agents and professional valuers working in the areas around Gatwick and Heathrow airports. This reliance on the opinions of estate agents, though it enabled a relatively rapid and widespread survey to be completed, is clearly less satisfactory than recording the actual selling prices of houses. The main results of the survey are shown in table 21.

**Table 21 The effect of aircraft noise on house prices near Gatwick and Heathrow airports: Roskill Survey**

| Percentage reduction in house prices<br>Class of property | Noise level (NNI)<br>35–45 | <br>45–55 | <br>Over 55 |
|---|---|---|---|
| Heathrow | | | |
| Low priced | 0 | 2·9 | 5·0 |
| Medium priced | 2·6 | 6·3 | 10·5 |
| High priced | 3·3 | 13·3 | 22·5 |
| Gatwick | | | |
| Low priced | 4·5 | 10·3 | — |
| Medium priced | 9·4 | 16·5 | — |
| High priced | 16·4 | 29·0 | — |

The average prices of houses in the low, medium and high categories were £3,000, £6,000 and £10,000 respectively. Whether or not the absolute Roskill figures can be accepted as reliable, there is no reason to doubt the evidence, which supports that of other surveys, that the proportionate decrease in house prices which is caused by noise increases sharply with the value of the property. It is also interesting to note that the estimated decrease in house prices for a given noise level was consistently higher around Gatwick than for houses near Heathrow. The area around Gatwick Airport is more rural and more middle-class than that around Heathrow, and the airport is less long-established.

A TRRL research paper contained what are claimed to be rough estimates of the costs and benefits of reductions of about 10 dB(A) in vehicle noise.[27] This estimate of the value of the benefits from noise reduction was based on the assumptions that a façade level (level of noise immediately outside a building) of 90 dB(A) will make a house uninhabitable; that a 60 dB(A) level is completely acceptable, involving no cost to the inhabitants; and that a linear relationship exists between the zero cost and 100 per cent cost noise levels. The 100 per cent cost for a 90 dB(A) façade traffic noise level was assumed to be equivalent to the value of the house rendered uninhabitable. The applications of values obtained from this linear relationship to the number of houses subjected to different levels of traffic noise produced a total noise reduction benefit of £14,200 million (or £6,750 million if a 65 dB(A) façade noise level is assumed to have zero cost)!

The obvious weakness of this calculation is the assumption of the linear relationship between house prices and the value of noise reductions. It is very difficult to believe that the value of the noise reductions is linearly related to house prices over all levels of traffic noise. Figures in the same TRRL paper show that the great majority of houses affected by traffic noise experience façade noise levels which are a long way below the 'destruction' level of 90 dB(A). A diagram of the percentage of houses exposed to different maximum façade noise levels does not show any houses with a façade level of more than 80 dB(A), and only 20 per cent of houses (out of all those affected by traffic noise) having façade levels of more than 70 dB(A) in 1970. Any marked non-linearity in the effect on the house-price/cost-of-noise relationship at the lower end of the range of noise levels would therefore have a very large effect on the estimate of the global value of noise reductions. Such evidence as exists suggests that the effect on the house-price/cost-of-noise relationship is markedly non-linear. Calculations based on the Roskill evidence on aircraft noise show that for 'high priced' houses the price reduction for houses in the 40–60 NNI range was approximately linear and could be expressed by the relationship

$$D = -35{\cdot}1 + 0{\cdot}96\ N$$

where D is the percentage fall in house prices (compared with an identical house in a noise-free area) and N is the Noise and Number Index. But if the NNI value for 100 per cent depreciation in house prices is calculated the value would be approximately 141. This is an impossibly high figure; houses would certainly be considered uninhabitable with a NNI figure of 70. In other words there must be marked non-linearity between the value of the noise-level/house-price reductions relationship at the level at which houses become uninhabitable and at those at which noise is a nuisance that can be borne. The issue is further complicated in that evidence in both the Roskill and Diffey surveys suggests that the noise-level/ house-price relationship differs for houses in different price ranges, and in different parts of the country. According to the Roskill evidence for example, the depreciation in house prices for houses near Heathrow with an average value of £3,000 suffering from noise conditions in the 45–55 NNI range was 2·9 per cent, while the equivalent figure for houses with an average value of £10,000 was about 13·3 per cent. The range in the Gatwick area for the same noise conditions and average prices was from 10·3 to 29·0 per cent.[21] Although in the absence of another indicator it is necessary to use the effect on house prices as a measurement of the value of reductions in noise levels, this is a somewhat unsatisfactory expedient. If some way could be found to make a direct measurement of what people would be willing to pay to obtain a reduction in traffic noise levels it seems unlikely that the resulting figure would vary directly with house prices. It would be surprising if the amounts people would pay to obtain a fall in the traffic noise (or the sums which they would require to be paid to them so that they were as well off with the traffic noise and the compensation as without either) varied directly with fluctuations in house prices. It would be dangerous to assume that the value of reducing traffic noise has increased in proportion to the recent rapid inflation in house prices, as the TRRL LR 441 model would imply.

The evidence which does exist about the money value which people put on reductions of traffic noise suggests that there may be important differences in the average values of people in different social classes. This is not, in the first place, a matter of income levels (although, of course, the uneven distribution of income does affect inter-group willingness to pay measurements) but represents a real difference in the evaluation of what is a mainly subjective cost. People who live and work with relatively high noise levels may find traffic noise less troublesome than those who generally enjoy quieter conditions. The complaints of countrymen who cannot sleep in the traffic noise of towns, and of town dwellers who are disturbed by farm noises in the countryside, may both be justified. There is a real danger that high middle-class values for the benefits of traffic noise reductions may be imposed upon the community when the real average values of what people would pay for quieter streets are much lower. This would be particularly unfortunate if, as is almost certainly the case, 'working-class' people are more likely to be exposed to traffic noise in their homes than are people living in middle-class housing areas. Of course, as the Council for the Protection of Rural England and the Civic Trust would no doubt point out, it is not only people in towns who suffer from traffic noise. But the relatively small number of country-dwellers exposed to high $L_{10}$ traffic noise would not have very much influence on the size of the total benefits from noise reductions even if their average valuation of benefits was high.

A recent attempt to measure people's willingness to pay for quietness was based on a survey of the installation of double glazing in council houses in an area of Hounslow affected by aircraft noise from Heathrow airport.[11] A grant-aided scheme for the installation of double glazing as a form of sound insulation was in operation in the period 1966–72 and the researchers were able to collect data on the installation of double glazing and to develop models which could be used to estimate a demand function. The type of double glazing installed was estimated to give a reduction of 14 dB(A) in internal noise levels, although this would presumably apply only when windows were closed. The average figures for willingness to pay estimated in this study were £45 for all bedrooms and £33 for living-rooms. Despite the considerable amount of effort which must have been expended in obtaining these estimates it cannot be said that they give any very reliable guide to the value of quietness. The figures are not related to particular noise levels, and no significant association was found between noise levels and readiness to install double glazing. Even if the whole area suffered from the same level of aircraft noise, the incidence of rail and road traffic noise may have varied between houses. Clearly what people are prepared to pay for a noise reduction from 74 to 60 dB(A) may have little relation to what they would pay for a reduction from 60 to 46 dB(A). Double glazing will not provide any appreciable sound insulation whenever windows must be opened for ventilation. Unless it is combined with air conditioning people may prefer noise to heat and lack of fresh air.

A serious objection to the use of expenditure on double glazing as a measurement of willingness to pay for quietness, which Starkie and Johnson acknowledged by stating that their estimates represented upper limits, is that double glazing also yields the important benefit of heat insulation. The prevention of heat loss and the reduction of fuel bills is normally the major 'selling point' for double glazing. Even if the main motive for installing double glazing in Hounslow was noise insulation householders must have been aware of the heat insulation benefits and the 'willingness to pay' measurement might be expected to include some payment for these benefits. Nevertheless, the Starkie and Johnson study is an interesting example of what the authors call the 'exclusion facilities' approach to the evaluation of environmental disbenefits and also of the very considerable research effort involved in producing just two average 'willingness to pay' measurements.

## Summary

*1. Noise is a subjective disbenefit. Transport noise reduces human happiness but it does not affect the output of goods and services and is not reflected in National Income statistics. The measurement of noise nuisance is difficult since individual reactions may differ considerably. There is some evidence that the degree of reaction to noise may be related to occupation or social class.*

*2. Transport noise is produced by road traffic, railways and aircraft. Road traffic noise tends to be continuous and is more complex than that of the other two modes. An increase in traffic speeds will tend to increase noise levels more than an equivalent percentage increase in vehicle flow. Lorries and buses contribute more to peak* ($L_{10}$) *noise levels than to overall levels. Greater noise reduction benefits can be obtained by reducing the noise of cars rather than that of 'lorries' on roads with vehicle flows of up to 400 per hour and with 20 per cent 'lorries' in the traffic composition. Reducing both car and 'lorry' noise will generally bring considerably greater benefits than reducing 'lorry' noise alone.*

*3. One of the most promising methods of reducing transport noise is by improvements in technology. Existing lorries, cars, buses and aircraft can be quietened to a limited extent, but major noise reductions involve redesigning engines and vehicles.*

*4. Where transport sound cannot be reduced at source, insulation by the use of sound barriers (for roads and railways) or by insulating houses can be an important way of reducing the nuisance from noise.*

*5. The separation of people from transport noise by distance is a most important remedial measure since other transport disbenefits such as pollution and accident causation may also be removed simultaneously.*

*6. Road vehicle noise and (to a limited extent) aircraft noise is controlled by government regulation. While it seems likely that regulation will continue to be necessary, some use of the price mechanism may be desirable in the future.*

*7. Knowledge of the costs and benefits of noise reduction measurements is at present quite inadequate as a guide to policy. It is quite impossible to determine whether, for example, a reduction of the maximum noise level for newly built lorries of less than 200 BHP from 89 dB(A) to 80 dB(A) would result in net gains or losses to the community. Similarly, the amount of compensation which should be paid to those affected by noise from new transport works under the provision of the Land Compensation Act of 1973 must be determined largely by guesswork.*

Chapter four

# Air pollution

## 1 Introduction

There is probably slightly less public concern about air pollution caused by transportation than there is about some other environmental costs of transport operation. In the TRRL 'High Street' Studies the 'effect of fumes on breathing' was ranked as less of a nuisance than delay in crossing the street, danger in crossing the street or noise (except in Putney where the effect of fumes had a slightly higher 'nuisance score' than 'danger in crossing the street').[1] There is no clear agreement about how far existing levels of transport-caused air pollution in Britain is injurious to health. But if the health of people living and working near heavily trafficked roads is being adversely affected by such pollutants as oxides of nitrogen or lead compounds then they are mostly unaware of this. The health hazard, if it exists, is at too low a level, and is too gradual in its effect, to be noticed. Dislike of air pollution relates mainly only to those forms of it which can be immediately smelled or seen. People are also generally aware that transport is not the only source of air pollution, but that there are other important contributors such as industry, waste disposal and (in some areas) domestic fires.

## 2 Sources and types of air pollution

The main atmospheric pollutants are carbon monoxide (CO); total suspended particulates, or solid particles of such substances as carbon, dust, and metal (TSP); oxides of sulphur, mainly sulphur dioxide ($SO_x$); hydrocarbons (HC); and oxides of nitrogen ($NO_x$). The sources of these pollutants in the U.S.A. and in West Germany are shown in tables 22 and 23.

**Table 22 Emission of air pollutants in U.S.A. (1970)**

| Source | CO | TSP | $SO_x$ | HC | $NO_x$ |
|---|---|---|---|---|---|
| | $10^6$ tonnes per year | | | | |
| Transportation | 112·8 | 0·7 | 1·0 | 19·8 | 11·9 |
| Fuel combustion in stationary sources | 0·8 | 6·9 | 26·9 | 0·6 | 10·2 |
| Industrial processes | 11·6 | 13·3 | 6·1 | 5·6 | 0·2 |
| Solid waste disposal | 7·3 | 1·4 | 0·1 | 2·0 | 0·4 |
| Miscellaneous | 17·1 | 3·4 | 0·3 | 7·2 | 0·4 |
| Total | 149·6 | 25·7 | 34·4 | 35·2 | 23·1 |
| Percentage change 1969–70 | −4·5 | −7·4 | 0 | 0 | +4.5 |

Source: *Environmental Bulletin*, Department of Trade and Industry, no. 13 (October 1972).

**Table 23 Emission of air pollutants in West Germany (1969/70)**

| Source | CO | TSP | $SO_x$ | HC | $NO_x$ |
|---|---|---|---|---|---|
| | $10^6$ tonnes per year | | | | |
| Transport | 8·0 | — | 0·1 | 1·0 | 0·9 |
| Industry | — | 0·8 | 0·3 | 0·9 | 0·2 |
| Power generation, heaters and incineration | — | 3·2 | 3·6 | 0·1 | 0·9 |
| Total | 8·0 | 4·0 | 4·0 | 2·0 | 2·0 |
| Estimated 1980 total | 8·0 | 2·0 | 4·3 | 3·5 | 4·0 |

Source: *Technology and the Environment*, Department of Trade and Industry reports from Scientific Counsellors, no. 2 (February 1972).

These tables show that transportation is a major contributor to emissions of carbon monoxide, hydrocarbons, and oxides of nitrogen. The relationship between cars and lorries in creating noise is, however, reversed for air pollution, cars being the main source of emission. The reasons for this are that petrol engines emit more pollutants that are a potential health hazard than do diesel engines (which are used in almost all heavy lorries), and that there are very many more cars than lorries on the roads. Table 24 gives a comparison of the composition of the exhaust gases of petrol and diesel engines in different running conditions.

**Table 24 Representative composition of exhaust gases**

| | Running condition | | | | | | | |
|---|---|---|---|---|---|---|---|---|
| | Idling | | Accelerating | | Cruising | | Deceleration | |
| Pollutant | Petrol | Diesel | Petrol | Diesel | Petrol | Diesel | Petrol | Diesel |
| CO | 69,000 | trace | 29,000 | 1,000 | 27,000 | trace | 39,000 | trace |
| HC | 5,300 | 400 | 1,600 | 200 | 1,000 | 100 | 1,000 | 300 |
| $NO_x$ | 30 | 60 | 1,020 | 350 | 650 | 240 | 20 | 30 |
| Aldehydes | 30 | 10 | 20 | 20 | 10 | 10 | 290 | 30 |

Source: P. Sherwood and P. Bowers, *Air Pollution from Road Traffic* (RRL Report LR352, cited in M. E. Burt, *Roads and the environment*, TRRL Report LR441).

These figures show that petrol engine exhaust emissions contain very much more carbon monoxide and hydrocarbons than diesel engine exhaust, more oxide of nitrogen, except when idling or decelerating (when the levels for both types of engine are very low), and more, or an equal quantity of, aldehydes. Estimates of the total annual emission of certain pollutants from petrol and diesel engines in Britain are shown in table 25.

**Table 25 Total estimated annual emissions of pollutants, diesel- and petrol-engined vehicles, Great Britain 1971**

| Pollutant | Estimated total annual emission (tons) Petrol engines | Diesel engines |
|---|---|---|
| CO | 5,000,000 | 80,000 |
| HC | 250,000 | 16,000 |
| $NO_x$ | 170,000 | 50,000 |
| Aldehydes | 10,000 | 3,000 |

Source: Report of Clean Air Society (cited in SMMT *Newsletter*, CV/8, September 1972).

In view of these figures it may seem surprising that lorries are widely accused of polluting the atmosphere and that Britain has legislation controlling smoke emission but does not, so far, place legal limits on the emission of much more dangerous pollutants from petrol engines. This is because diesel engines can, in some circumstances, emit black smoke, consisting of unburnt particles of carbon. These emissions of black smoke do not make any significant contribution to atmospheric pollution, so do not appear in the measurements of pollutants. But unlike other gaseous emissions, they can be seen, and may smell unpleasant, so that far more attention has been paid to smoke emission in Britain than to the potentially much more dangerous contents of vehicle exhausts. A paper on techniques for reducing exhaust emissions states that 'Diesel smoke is offensive to the senses and, although it presents a road safety hazard through impairing visibility, public concern over it stems mainly from the ease with which it can be recognised compared with gaseous emissions'.[6] The carbon particles which form black smoke are not considered to be harmful to health at roadside concentrations. The unpleasant smell of diesel smoke is probably caused by unburnt hydrocarbons though this has not been proved conclusively.[10]

The effect of oxides of nitrogen on health at low concentrations has not been fully researched. Nitric oxide can act as a chemical asphyxiant and nitrogen peroxide is an irritant when inhaled and can lead to respiratory failure. The most generally accepted maximum level for exposure to $NO_2$ is 5 ppm (parts per million).[10] Typical busy street concentrations of $NO_x$ are about 0·2 ppm.[4] It is possible that oxides of nitrogen may have a long-term effect on health and thus may be potentially more dangerous than carbon monoxide.[4] Hydrocarbons may have a photochemical reaction with oxide of nitrogen in strong sunlight conditions to produce smog which irritates the tissues of the eyes, nose and throat. It is believed that the probability of the right meteorological conditions occurring to cause smog in Britain are very low.[10] The possibility of lead pollution in the atmosphere has now given rise to some concern but this is caused by additives to petrol and is not a problem with diesel engines.

Carbon monoxide is very poisonous in high concentrations, and even at low concentrations it can impair mental ability and could lead to road accidents if people's alertness was reduced. There is some disagreement about the maximum levels of concentration which should be allowed if all health hazards are to be avoided. It has been suggested that the maximum acceptable level should be 50 ppm,[7] but some research workers believe that 30 ppm should be the maximum level while others have argued for an even lower level of 9–10 ppm. The concentration of carbon monoxide in a typical fairly busy street in Britain would be in the range 10–20 ppm,[7] but considerably higher figures have been recorded. Measurements at the kerbside on London Road in Reading for the periods 27 May–15 July and 7–23 December (1971) showed that the carbon monoxide concentration exceeded 10 ppm for more than 7 per cent of the summer monitoring period and for about 25 per cent of the winter period.[8] The maximum concentration recorded was 77 ppm. The total number of minutes per week when concentrations exceeded 50 ppm ranged from 0·5 to 20·9. The health hazard from carbon monoxide produced by road transportation has been put into perspective by

Professor Lawther of the British Medical Research Council who claims that a cigarette smoker inhales about seven times as much carbon monoxide with every puff as the highest concentration found in heavy traffic.[9]

Lead pollution is of two forms, particulates (or small solid particles) and volatile lead compounds. Again the extent of the health hazard is not fully agreed but the main fear is that dangerous concentrations of lead can build up gradually in the human body. The average concentration of particulate lead found in the Reading study (excluding weekend figures) was 4 $\mu g/m^3$ (micrograms per cubic metre) which can be compared with the concentration at a site remote from traffic flows of 0·25 $\mu g/m^3$.[8] Another study of total lead pollution was made at four sites near busy roads. The average level at two filling stations was 2 $\mu g/m^3$; the lead level in a road tunnel averaged 15 $\mu g/m^3$ with a range from 5 to 28 $\mu g/m^3$; while on a motorway the average level was 10 $\mu g/m^3$ with a range from 8 to 20 $\mu g/m^3$.[11]

## 3 Controls on transport-caused air pollution

At present it is only smoke emission and the level of lead in petrol which are controlled by legislation in Britain. Smoke emission is now controlled by regulations relating to vehicles manufactured on or after 1 October 1972 and first used after 1 April 1973. The test consists of measuring the obscuration to light caused by a sample column of exhaust smoke. Existing tests relate to design standards of engine types but there are proposals to make the standards apply to all engines actually manufactured.[12]

It was announced by the Secretary of State for the Environment in 1972 that the permitted level of lead in petrol would be progressively reduced from 0·84 g/l (grams per litre) to 0·45 g/l. Lead is added to influence the combustion process and reduce 'knock'. According to the Automobile Association study, total elimination of lead from petrol would 'increase the likelihood of premature engine failure very considerably'.[9] There have been proposals in West Germany to reduce the lead content in petrol to zero by 1980.[13]

Legislation on the emission of carbon monoxide, hydrocarbons, and oxides of nitrogen has been mainly confined to the United States. The U.S. Clean Air Act of 1970 gave powers to the American government to control various forms of air pollution, including those caused by transportation. Individual states have also made their own regulations and the standards adopted in California have had considerable influence. There have been various proposals for making the U.S. regulations more stringent but their implementation has been delayed partly to meet the objections of vehicle manufacturers and, more recently, because of the possible adverse effect on fuel consumption.[14] There is no doubt that the most severe standards proposed for the U.S. would involve heavy costs for both diesel and petrol engined vehicles. One estimate made by the U.S. Environmental Protection Agency suggested that the U.S. would have to pay $8,000 million annually to meet the requirements of the Clean Air Act up to 1979 and that car emission controls would account for more than half of this figure.[15] The figures in table 26 give one version of a proposed more severe U.S. standard, originally intended to come into operation in 1976, compared with some EEC proposals and the actual emissions of typical European cars in 'good tune'.

**Table 26 Some proposed maximum emission standards and typical existing emission levels**

| | Exhaust emission, gms per mile | | |
|---|---|---|---|
| | HC | CO | $NO_x$ |
| U.S. Federal proposed | 0·46 | 4·7 | 0·39 |
| EEC proposed | 4·5 | 55 | Not specified |
| Typical European car | 6·0 | 80 | 3·5 |

Source: M. Jacobson, *Air Pollution and the Motor Vehicle* (Automobile Association memorandum, October 1971).

## 4 Methods of reducing pollution from vehicle emissions

The emission of black smoke occurs when carbon in the fuel does not burn completely, because there is insufficient oxygen available. In normal burning all carbon should combine with oxygen to form carbon dioxide, which is harmless (and which, with oxygen, makes up about 18 per cent of emission from the exhaust of a diesel engine).[6] The basic cause of smoke is therefore over-fuelling, or a too high ratio of fuel to air. All diesel engines are fitted with a stop to limit the quantity of fuel used and this should ensure that the exhaust emissions are invisible. Smoking occurs when the maximum fuel stop is worn or wrongly adjusted, when there is a fault in the fuel-injection system, or when lorries are starting from cold and an excess fuel device allows more than the normal maximum amount of fuel to be burned.[14]

Engine and vehicle manufacturers interviewed during the *Living with the Lorry* survey were generally not greatly concerned about the problem of meeting existing legislation standards. All those producing engines said that they could meet the British smoke-control standards. Manufacturers who had investigated the issue also believed that they could meet the Californian standards relating to the emission of carbon monoxide, hydrocarbons and oxide of nitrogen. The Californian maximum emission levels were:

CO =40 grams per BHP/hour
$NO_x$, HC } = 5 grams per BHP/hour

Typical comments by engine manufacturers interviewed were that 'the engines would meet the Californian standards without using scrubbers, filters or other devices outside the engine' (Cummins); 'engines would be manufactured to meet U.S. requirements, but the new 1975 regulation emission level would be expensive to achieve, and the durability of the measures taken to reduce emission would be uncertain' (Perkins). Mr R. Elliot of Perkins Engines pointed out that $NO_x$ emissions from diesel engines tend to be low because of the low speed of vehicles. The findings in an article by John Dickson-Simpson on 'Making diesels sociable' were very similar to those of our survey.[16] He also reported tests on a V8 engine which would normally be turbocharged and which (in independent tests on the naturally aspirated form) yielded CO emission of only 3·5 grams/BHP/hour and $NO_x$ of 8·3 grams/BHP/hour.[16]

Opinions on how to reduce emissions of $NO_x$ and the other pollutants differed amongst those interviewed. Neither Cummins nor Perkins favoured the use of scrubbers or catalytic devices. Mr Elliot did not think that these devices could cope satisfactorily with $NO_x$ emissions. Mr D. Hughes of Rolls Royce, on the other hand, considered that wet scrubbers, which are already used satisfactorily on diesel engines in trucks and dumpers working underground, could be adapted for use on trucks of 300 BHP. These scrubbers would cost about £100 each at 1972 prices. Vehicle manufacturers interviewed were generally unhappy about any device which would add to the unladen weight of vehicles and so reduce payload.

A major technological problem in dealing with emissions is that reducing the emission of one substance may increase that of another. Thus reducing the fuel/air ratio will cut smoke levels but may increase the proportion of unburnt hydrocarbons and oxides of nitrogen.[6] Retardation of fuel injection will reduce gaseous emissions but may increase the amount of smoke. Turbocharging will reduce smoke emission over most of the ranges of operating speeds, and, according to Messrs Arbury and Morrell of Holset Engineering, the problem of smoke emission when turbocharged vehicles accelerate from rest or from very low speeds has now been largely overcome. Turbocharging appears to be a more acceptable way than decreasing the fuel/air ratio of reducing smoke emission because it increases power and thermal efficiency, whereas decreasing the fuel/air ratio will also reduce engine power. Turbocharging does not reduce oxide of nitrogen emission, however, and may even lead to an increase.[6] Another important approach to the emission reduction problem is to vary the composition of the fuel used. The quantity of unburnt hydrocarbons at idle or high load (of the engine) can be reduced by using fuel of good ignition quality.[6, 17]

Methods of reducing exhaust emission may not only have opposite effects on different pollutant levels but may conflict with some other environmental aim. Reducing fuel/air ratios, and thus reducing engine power, would reduce the power-to-weight ratios of vehicles, and thus tend to increase delays to other road users. The advance of injection timing will reduce smoke levels (though increasing gaseous emissions) but this will tend to increase noise levels. Various emission reduction techniques will have an adverse effect on fuel consumption (which is a matter of relevance to environmental as well as economic considerations) and vehicle 'drive-ability'.[17]

As has already been suggested the problem of meeting rigorous emission controls is generally more severe for car manufacturers than it is for firms producing diesel-engined lorries. There are a number of possible ways in which petrol engine technology could be adapted to reduce gaseous emissions. These include exhaust gas recirculation and changes in the fuel/air ratio, combined with various forms of thermal or catalytic oxidation. Some of the methods are basically 'preventive' (preventing the production of unwanted emission) while others are 'curative' (removing the unwanted gases from exhaust emissions) but the two approaches are usually combined.[17] The choice of an optimum technique is very complicated. Costs and reliability differ considerably. Some techniques require, for example, a very accurate control of mixture strength, which normal carburetter systems may not be able to supply. The effect of different control methods on the reduction of different emissions will also vary. Thus the use of a very rich air/petrol mixture would reduce $NO_x$ emission considerably but cause a large increase in the emission of hydrocarbons and carbon monoxide; these would then have to be removed by thermal or catalytic oxidation.[17] As with changes in diesel engine technology there may be other losses to offset the gains from reduced gaseous emission. All the techniques for reducing gaseous emissions have some adverse effects on vehicle power, fuel consumption or durability, and in many cases on all three.[17] The Automobile Association memorandum stressed the importance of proper vehicle maintenance and particularly of correct ignition timing as means of minimizing unwanted gaseous emissions from existing cars.[9]

Diesel-engined rail trains produce gaseous emissions as do lorries, but the problem is much less serious since rail traffic density is usually relatively very low. Nevertheless there has been some concern in the U.S. about the contribution made to air pollution in urban areas by diesel trains.

## Summary

*1. The benefits which would result from reducing gaseous and other unwanted emissions from vehicle exhausts are even more difficult to evaluate than those associated with noise reduction. The long-term effects on health of air pollution caused by transportation are still largely unknown.*

*2. In Britain, legislation controls the emission of visible smoke from diesel engines which is unpleasant but not believed to be a health hazard.*

*3. Petrol engines produce most of the transport-caused air pollution in Britain (and are entirely responsible for transport-caused lead pollution).*

*4. It is technically possible to reduce gaseous and lead emissions from petrol and diesel engines but the costs involved in meeting the most stringent standards proposed in the U.S. would be very high. These costs would take the form not only of increases in the capital and operating costs of vehicles but would also include the costs of reduced vehicle power (with slower journey times), higher fuel consumption and poorer vehicle 'durability'.*

*5. The emissions from existing vehicles can be reduced with good maintenance and carefully adjusted ignition timings.*

*6. Until more is known of the health effects of transport-caused air pollution it is impossible to say whether the adoption of severe maximum emission controls is, or is not, justified. However it seems likely that there would be net benefits from introducing some controls on gaseous emissions in Britain and in other EEC countries.*

Chapter five

# Road wear and track costs

## 1 Introduction

Transport may be regarded as consisting of three basic elements: a means of locomotion and carriage; terminal facilities; and a track over which movement can take place. 'Track cost' is generally taken to mean the total cost of providing, maintaining, and policing the public highway system. Where a natural highway exists as in the case of air and sea transport, track cost is negligible. For road and rail, track cost is significant, as we see in table 27.

**Table 27 Road and rail expenditure in the U.K., 1972–3**

| Road | (£m) | Rail | (£m) |
|---|---|---|---|
| New construction and improvements | 596·4 | Permanent way | 82·3 |
| Maintenance | 304·5 | Track maintenance | 16·8 |
| Administration | 80·8 | Signalling | 52·0 |
| | | General expenses | 121·1 |
| | 981·7 | | 272·2 |

Source: *Road expenditure in Great Britain* (HMSO, 1974); British Railways Annual Accounts, 1974.

The problem of track costs has long been a central issue in transport economics. There has been much debate about the identification and calculation of track costs, and whether different transport users make an efficient and equitable contribution towards their track costs. In recent years, however, there has been a shift of emphasis away from the track cost problem towards assessing the effects of different forms of transport on the quality of the environment. The extent to which heavier commercial vehicles tend to wear out the road track more quickly than lighter vehicles is an important issue in considering the problem of allocating road infrastructure and maintenance costs amongst different classes of road users, but the environmental implications are not of any great consequence. Heavier vehicles may be rejected because they cause more noise, or are believed to cause more accidents, but changes in the technique of road construction do not appear to have any obvious environmental significance. Improving the road surface to carry heavier axle loads may be compared to strengthening rail track to carry the advanced passenger train, a procedure to which there are unlikely to be any environmental objections. Increasing road wear caused by heavier vehicles could have some environmental implications insofar as other road users suffer when roads become bumpy, or are closed more frequently for repair, and all suffer if the countryside is disfigured by the extraction of materials for road construction and maintenance.

Although there is little direct relationship between them, track costs and environmental costs share many similar characteristics. A closer examination of the track cost problem, and possible solutions, may throw light on the issues arising in tackling the wider environmental costs of transport. The basic public policy problem arises because road users do not have to pay any charge that directly reflects the road damage which they cause. British Rail can include infrastructure costs in transport charges. For roads and inland waterways, however, the ownership and control of the infrastructure is largely separate from that of transport services, and so the danger arises that transport users will not take into account the burden they impose on the public purse by track wear and tear. In this sense track cost is yet another element, along with congestion and other externalities such as noise, smoke and so on, in the marginal social cost, defined as the cost to society of the last additional unit of traffic to use the particular infrastructure facility. The railways have long complained of unequal competition and claimed that road as well as rail should meet track costs, though large governmental subsidies to British Rail take the edge off their argument that they have been seriously disadvantaged.

Figure 1 postulates a hypothetical relationship between marginal private cost and marginal social cost for a range of vehicle mileages. As vehicle mileage increases, operating costs rise, so the marginal private cost curve is upward sloping. The marginal social cost curve is derived from the marginal private cost schedule by adding to it increased marginal road wear and tear cost, marginal social cost of congestion, and the marginal cost of other externalities such as noise, smoke etc.[1] The provision of road track is a classic example of a public good and this alone would justify public intervention. Moreover, each private road user would tend to put a low value on the advantage he would receive from track improvement, to minimize his share of the cost. Again, private contracts between the millions of actual and potential road users, to share the cost of track, are clearly an impossibility.

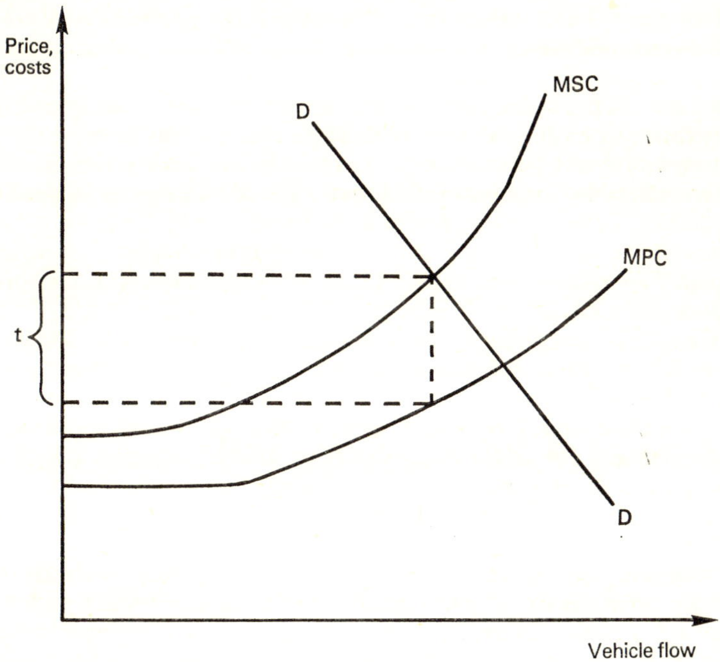


*Figure 1. Taxation, and marginal social and marginal private cost. (D = demand, MSC = marginal social cost, MPC = marginal private cost, t = tax.)*

The existence of major indivisibilities in the provision of track creates further problems. The Department of the Environment estimates the cost of building a dual three-lane by-pass as only 25 per cent above that of a dual two-lane by-pass, so that if widening is required it is usual to provide an extra lane. Walters points out that where existing roads are being improved to four lanes there is little difference to cost whether the existing road is two or three lanes wide.[1]

The characteristic of joint costs complicates any attempt to attribute costs of track to categories of individual users. One proposal is that each category should at least bear the extra cost of providing for it.[2] However, the size of the extra construction costs on new roads and improvements to existing ones necessary for the heaviest axles for example, depends on the assumption that some road track has already been provided. The same problem arises in the case of vehicle clearances. Only the tallest vehicles need the highest clearance, and it is appropriate to charge them this increment. What is not clear is the share they should have in the lesser increments which are necessary for them, and for smaller vehicles. Given these difficulties it is suggested that there is 'little economic meaning to share these total costs out between vehicles, since the relativities obtained could not be said to reflect true relative costs even in principle. There is, it is therefore agreed, a strong case for concentrating on the short run costs alone'.[3] There may be more choice between different types of road surface. It has been estimated that whereas flexible pavements would have a life of 11–23 years (depending on details of construction), concrete pavements, even in motorway slow-lane conditions, would last for 105 years before requiring structural repair.[4] To some extent it may be possible to trade-off maintenance costs against construction costs.

Given the characteristics of track cost some public intervention is essential, but the form such intervention should take is by no means clear. The static theory of optimum resource allocation does not tell us which public policy to adopt. Even assuming that an omnipotent policy-maker knew all the utility functions, all the production functions, and all the technological functions relating vehicle operation to road wear and tear, and given no undesirable effects on distribution of income, the policy-maker could still follow either of two policies which would be equally efficient. He could allocate resources by regulation, simply telling each vehicle operator what his track wear and tear should be. Alternatively, he could use the appropriate marginal conditions to set fees or taxes for road wear and tear, and then allow market forces to allocate resources, along with prices of other goods set by competitive markets. In principle the policy-maker would require exactly the same information to carry out the two policies.

In the real, and very imperfect world, policy-makers in the U.K. have adopted a mixture of regulations and taxes to tackle the problem of track cost. The situation is further complicated by the fact that the policy-makers have used a single instrument to achieve a variety of results. Thus regulations for vehicle construction and use are drawn up not only in relation to track cost, but also to many other factors, including public safety. Similarly, it could be argued that vehicle taxation is determined by revenue objectives rather than track cost considerations.

The economist can quantify the costs and benefits of regulations, but the actual formulation of regulations is largely a technical matter. In the case of prices, however, the economist may make a more direct contribution. Given the measurement problems for many transport externalities the only currently feasible policies are regulations. In the case of track costs, however, there is more scope for using the price mechanism to make road users pay for road wear. This Chapter examines the existing structure of taxation on commercial vehicles, and attempts to formulate a system which will more accurately reflect the track cost burden they impose on the public purse. The importance of this exercise is increased because it is the aim of the EEC to devise such a charging system.

## 2 Objectives of a charging system

The first stage in developing a road user tax must be to formulate the objectives of a charging system for commercial vehicles. It could be argued that a charging system should have four major objectives:

(i) It should facilitate the best possible use of existing track facilities.
(ii) Charges may need to meet a budgetary requirement, to cover the costs of expanding the track network, and/or to meet an Exchequer general revenue target.
(iii) The charging system should be regarded as equitable.
(iv) The administrative costs should not exceed the benefits of the charging system.

Apart from a number of minor fees and charges, taxation is currently raised from transport operators in three main forms: fuel tax, vehicle excise duties, and value added tax. Since the latter is refunded, any charging system for road wear and tear must be based on fuel tax and/or vehicle excise duties. Fuel tax has three characteristics which make it particularly suitable for a charging system. It is paid with each vehicle kilometre travelled, it varies with the nature of the vehicle, and with the speeds at which vehicles travel. The excise licence duty is an annual payment and is related to a vehicle's unladen weight. It rises from £40 for vans not exceeding 16 cwt unladen weight to £683·95 for vehicles of 9 tons 15 cwt to 10 tons used with a trailer. For vehicles above 4 tons unladen weight the duty rises by £18 for every 5 cwt increase in unladen weight. Both taxes suffer from severe deficiencies. They do not adequately reflect variations in the costs of road provision from one part of the road system to another, nor variations in the level of congestion costs from one part of the system to another, and between time of day, nor variations in the costs imposed, nor the benefits received by particular categories of vehicle.

As well as being deficient from an 'efficiency' point of view, vehicle taxation may also be considered inequitable. The most authoritative analysis of expenditure on track is still the 1968 *Road Track Cost Report* which uses estimates for 1965–6. A more recent study[5] provides comparable figures for 1970–1 (see table 28). For both years a simple accounting approach of comparing revenue and expenditure indicates a large surplus. The total revenue collected from fuel tax and vehicle licence duty in 1965–6 and 1970–1 amounted to £926 million and £1,583 million respectively.

**Table 28 Public road costs**

| | 1965–6 | | 1970–1 | |
|---|---|---|---|---|
| | Current expenditure approach (£m) | Public enterprise approach (£m) | Current expenditure approach (£m) | Public enterprise approach (£m) |
| Capital: | | | | |
| New roads and major improvements | 179 | — | 510 | — |
| Minor improvements | 36 | — | | — |
| Capital charges | — | 390 | — | 728 |
| Maintenance | 96 | 96 | 155 | 155 |
| Cleansing | 23 | 23 | 33 | 33 |
| Lighting | 9 | 9 | 15 | 15 |
| Policing | 52 | 52 | 75 | 75 |
| Administrative | 40 | 40 | 64 | 64 |
| Accidents (over and above those covered by insurance) | 15 | 15 | 18 | 18 |
| | 450 | 625 | 870 | 1,088 |

Source: *The Road Track Cost Report* (HMSO, 1968): 'Transport policy – roads and road finance', *Socialist Commentary* (April 1975).

A major problem in estimating expenditure on roads is the treatment of capital. No accounts on a commercial basis have been kept for roads, so no conventions have come to be accepted for presenting to current users the fact that they are enjoying benefits from past streams of expenditure, often going back centuries. Of the many possible solutions, two approaches were adopted. The Current Expenditure basis includes capital and other costs actually incurred on roads in a given year, and this gives an excess of revenue of £476 million and £713 million in 1965–6 and 1970–1 respectively. The Public Enterprise basis is more of a hypothetical estimate. It is a calculation of what might be the capital charge for roads if they had in fact been organized like a nationalized industry. This approach gives an excess of revenue of £301 million and £495 million in 1965–6 and 1970–1 respectively. If more sophisticated economic pricing criteria are used to estimate revenue there is still a significant excess over expenditure. The Road Track Cost Study employed two such methods. First, short-run marginal cost pricing, which covers only those costs which cannot be escaped in producing an additional unit of output, produces a surplus of revenue of £734 million in 1965–6, and £1,223 million in 1970–1. Second, long-run marginal cost pricing, which attempts to cover all future expenditures on the roads, both capital and running costs, would produce a surplus of revenue similar to the public enterprise approach.

Whatever the approach adopted, in both 1965–6 and 1970–1 the revenue collected from road user taxation greatly exceeds the burden on the public purse of providing track. It is often argued that this excess of tax revenue is both equitable and efficient because it reflects other elements of marginal social cost. The standard Pigovian argument would be that vehicle taxation internalizes not only road wear and tear, but also the environmental costs of congestion, and makes allowance for the scarcity value of foreign exchange expended on fuel imports. In figure 1 the optimal level of user taxation is given by 't', where the willingness to pay for the marginal vehicle kilometre travelled equals the aggregated constituents of marginal social cost.

It is certainly true that since the quadrupling in the price of oil by the oil producers, fuel is an exceptionally scarce and valuable commodity. The U.K. balance of payments deficit on current account in 1974 was £3,668 million and of this sum £3,423 million was due to oil imports. The balance of payments deficit reflects the fact that the real value of savings in fuel costs are even higher than the face value of the financial savings. The current market price of fuel in the U.K. does not reflect its true social marginal opportunity cost. It would of course be sheer coincidence, if not miraculous, if having deducted the other elements of marginal social cost, the remainder of the excess of tax revenue over track expenditure reflected the real social value of fuel.

Vehicles undoubtedly generate significant environmental costs in the form of noise, smoke emission, visual intrusion and severance. However, as we have seen, there is no consensus on a money measure of environmental nuisances created by traffic, nor even consensus on a methodology for deriving money values.[6] An environmental tax, while in principle laudable, in practice is beset with problems. It would be wishful thinking, bordering on stupidity, to think that existing vehicle tax included in any rationally thought-out, efficient or equitable way, environmental costs imposed by vehicles.

One justification for the excess of tax revenue over road expenditure is that it may be regarded as a contribution to general taxation. Clearly there is no reason why road users should not be taxed, but it is difficult to see why they should be singled out for specially high rates of taxation, particularly when compared with other modes of transport. Thus for inter-urban traffic, assuming there has been no over-investment in the inter-urban road system, tax revenues should substantially exceed the relevant costs. By contrast, the railways and air transport pay negligible taxes, and have consistently failed to cover long-run marginal cost. Consequently the existing system of taxation could seriously distort the pattern of demand, and hence resource allocation. The same kind of distortion could occur in regard to urban areas. If congestion costs are included as a constituent of marginal social cost but excluded from the charging system, then tax revenue is likely to be less than costs, and the real costs of car use in urban areas under-estimated.

An examination of how the existing tax is allocated between different types of vehicle reveals a situation that is neither efficient, nor equitable. The *Road Track Cost Report* adopted as its criterion for efficiency and equity that the ratio of the revenue raised from vehicles of different types, to the costs imposed by those types, should be the same. The costs covered included those varying with vehicle weight, and other costs such as cleansing, lighting, policing, and administration, which were allocated on a mileage basis. Since heavier vehicles generally have a higher annual vehicle mileage than lighter vehicles they attracted a larger share of these costs than if allocations had been proportioned to the numbers licensed. But the extra cost allocation was a fraction of mileage, not of weight. No allowance was made for congestion, or environmental factors such as noise and air pollution. Capital was included on the 'current expenditure' rather than the 'public enterprise' basis. The Ministry Report found that for 1965–6 revenue exceeded costs for all classes of vehicle (see table 29). A more recent study for 1970–1 shows a marginal reduction in the ratio for cars, and heavy goods vehicles, a significant reduction for light vans, and for public service vehicles a reversal, with cost exceeding revenue.

The revenue/cost ratios for goods vehicles of different unladen

**Table 29 Revenue/cost ratios by vehicle class**

| Vehicle type | Revenue/cost 1965–6 | Revenue/cost 1970–1 |
|---|---|---|
| Cars | 2·1:1 | 2·0:1 |
| Public service vehicles, | 1·4:1 | 0·9:1 |
| Light vans | 3·3:1 | 2·1:1 |
| Heavy goods vehicles | 1·8:1 | 1·5:1 |

Source: *The Road Track Cost Report* (HMSO, 1968); 'Transport policy – roads and road finance', *Socialist Commentary* (April 1975).

weights were also estimated in the *Road Track Cost Report* for 1965–6. For every category revenue exceeded cost (see table 30). The study for 1970–1 shows a marginal reduction in revenue as against cost, but for every category revenue still exceeded cost. The results are averages and within each category some vehicles will add more to cost than to revenue. More seriously, this efficiency criterion may only be useful in comparisons between categories, where the main source of competition for a particular vehicle category is another vehicle category. Where intermodal substitution is important, for example heavy goods vehicles and rail, this criterion is not applicable. Furthermore, the road track costs imposed by vehicles do not depend primarily on their weight. The unladen weight of a goods vehicle is an inadequate guide to the load it imposes, as shown in table 31. A combination of distance travelled laden and unladen, and the number and juxtaposition of axles determines road wear and tear.

**Table 30 Revenue/cost ratios for heavy goods vehicles**

| Unladen weight (tons) | Revenue/cost 1965–6 | Revenue/cost 1970–1 |
|---|---|---|
| 1½–2 | 2·1:1 | 2·2:1 |
| 2–2½ | 2·7:1 } | 2·1:1 |
| 2½–3 | 2·5:1 } | |
| 3–4 | 1·7:1 | |
| 4–5 | 1·7:1 | 1·4:1 |
| 5–6 | 1·7:1 | 1·4:1 |
| 6–7 | 1·5:1 | 1·5:1 |
| 7–8 | 1·6:1 | 1·4:1 |
| 8+ | 1·6:1 | 1·4:1 |

Source: *The Road Track Cost Report* (HMSO, 1968); 'Transport policy – roads and road finance', *Socialist Commentary* (April 1975).

Given these shortcomings an attempt has been made to revise taxation for commercial vehicles, so as to devise a charging system which will reflect more accurately at least one constituent of marginal social cost, namely road wear and tear.

**Table 31 Average axle loadings of heavy goods vehicles**

| Unladen weight (tons) | Number of axles 2 (tons per axle) | 3 | 4 |
|---|---|---|---|
| 3–4 | 4·2 (8·4) | — | — |
| 4–5 | 6·2(12·4) | 4·6(13·8) | — |
| 5–6 | 6·1(12·2) | 5·0(15·0) | — |
| 6–7 | — | 5·4(16·2) | 5·0(20·0) |
| 7–8 | — | 6·1(18·3) | 5·1(20·4) |
| 8–9 | — | 7·2(21·6) | 5·2(20·8) |
| 9+ | — | — | 5·8(23·2) |

(Figures in brackets relate to average gross vehicle weight)

Source: *The Road Track Cost Report* (HMSO, 1968).

## 3 A new road wear tax

Fuel tax, as we have noted, has many advantages as a basis for a track cost charging system. In particular, it is paid per vehicle kilometre travelled. Unfortunately it would be very difficult to make this reflect the different contributions of vehicles to track costs. A possibility is to have different coloured fuel for vehicle categories according to their track cost effects, but this would create obvious monitoring problems. *Faut de mieux*, it is the vehicle's excise licence duty which is refined.

Three separate charging systems have been devised corresponding to three policy objectives:

(i) A charging system to reflect only short-run marginal track maintenance costs of vehicle operation.
(ii) A charging system to reflect capital expenditure on road track and short-run marginal track maintenance costs of vehicle operation.
(iii) A charging system to reflect capital expenditure on road track, short-run marginal track maintenance of vehicle operation, and a budgetary constraint set by the Exchequer.

*Policy objective* (*i*)

Having decided to provide a certain volume of output, and with capital invested, the only resources affected by a decision to produce an extra unit of output, for example a vehicle kilometre, are those which vary directly with the level of output. If charges include costs of resources committed in such a way that the resources cannot be released quickly for other uses, then users may be led to choose a service or a product which is more expensive in terms of current resources used, and optimal resource allocation will not be achieved. Evidence on short-run marginal track maintenance costs attributable to different categories of commercial vehicles is limited. Some tentative estimates are presented in the Ministry of Transport Memorandum to the Geddes Committee on licensing.[7] Maintenance consists of a wide range of activities which are not all related in the same way to use of the roads by particular vehicle types. Following the Geddes submission four categories may be distinguished:

(a) Repair
(b) Preventative maintenance
(c) Miscellaneous work on drainage, kerb renewals, etc.
(d) Provision of signs, signals and road markings

One of the Ministry's estimates (the least favourable to commercial vehicles) suggests that 35·4 per cent of total road maintenance expenditure should be allocated to vehicles over 1 ton unladen weight. The total 1972–3 maintenance expenditure on roads in Great Britain was £231 million, and 35·4 per cent of this figure is £81·8 million.[8]

It is undoubtedly the case that road wear and tear increases with axle-loadings, but there has been considerable controversy about the nature of the relationship. The most thorough experimental work carried out measuring the effect of different axle-loadings on road wear was the famous AASHO (American Association of State Highway Officials) tests. These tests were carried out over a period of two years in Illinois, on six tracks, with help from the U.S. army, and at a cost of $27 million. The main conclusion has been developed into a formula stating that relative road wear is a function of the fourth power of the ratio of axle-loadings. This means that virtually the whole of road 'wear and tear' maintenance costs should be attributed to heavy vehicles. Criticism of the AASHO findings has been centred around their application to British and European conditions. A detailed criticism was made by a team commissioned by the International Road Transport Union,[9] who argued that the large number of parameters in the AASHO study made it difficult to isolate the influence of a single variable such as axle-loading. Even with so many parameters, however, it was argued that allowance was not made in the tests for the effect on road wear of the nature of the subsoil, for long-run climate and weather conditions, for the effect of maintenance, or for the mixture of traffic using the road. The team concluded that the AASHO results were not applicable to operating conditions on European roads. The relevance of the AASHO findings to British conditions has also been questioned by the Ministry of Transport. In evidence submitted to the Geddes Committee on Carriers' Licensing, the Ministry argued that the effect of frost was much more important in Illinois than it would be in British conditions, and that the concentration of heavy axle-loadings in the test was essentially different from the mixed light and heavy traffic actually experienced on British roads. Thus it was argued that the 'fourth power' rule was harsh in its treatment of heavy axle loadings in Britain. Nevertheless, in the absence of any other authoritative empirical work, the 'fourth power' formula is generally employed in assessing road wear and tear.

The road 'wear and tear' effect for each category of gross vehicle weight is derived from the estimated number of vehicles in each group, kilometres travelled laden and unladen, and number of average standard axles. The number of standard axles is estimated by assuming that a single pass over a section of road by an 18,000 lb axle has a road damage factor of 1·0, and all other axle weights are related to it on the basis of the 'fourth power' law.[10] Road maintenance expenditure attributable to commercial vehicles in 1972–3 is then allocated to each category on the basis of the following formula, and the results shown in table 32 on p. 100.

Road 'wear and tear' cost arising from vehicles in category i

$$= R \left[ \frac{A_{li}K_{li}V_i + A_{ui}K_{ui}V_i}{\sum_{i=1}^{56} (A_{li}K_{li}V_i + A_{ul}K_{ui}V_i)} \right]$$

Road 'wear and tear' cost per vehicle in category i

$$= R \left[ \frac{A_{li}K_{li} + A_{ui}K_{ui}}{\sum_{i=1}^{56} (A_{li}K_{li}V_i + A_{ui}K_{ui}V_i)} \right]$$

where: $R$ = Aggregate road maintenance expenditure attributable to commercial vehicles

$A_{li}$ = Estimated standard axle laden

$A_{ui}$ = Estimated standard axle unladen

$K_{li}$ = Annual average kilometres laden

$K_{ul}$ = Annual average kilometres unladen

$V_i$ = Number of vehicles in each of the 56 gross vehicle weight categories distinguished

As one would expect from the application of the 'fourth power' law, a tax aimed at internalizing road wear and tear would have a much greater impact on the heavier vehicles. A two-axle rigid lorry with a gross weight of 5 tons would pay only £1, as against £685 for a 32-ton four-axle articulated vehicle. However, the number of axles is a crucial factor, as illustrated by the fact that with the addition of one more axle to a 32-ton lorry the estimated road wear caused by the vehicle would be reduced by more than half, and a road 'wear and tear' tax would fall from £685·00 to £317·00. Again, a 24-ton four-axle articulated unit would pay £146·50, whereas a 24-ton three-axle rigid would pay £414·00.

It is sometimes suggested that the existing vehicle excise duty may be regarded as a charging system for road 'wear and tear'. The charge does indeed rise with unladen weight, but a linear relationship, (where $X > 16$):

$$L = -81 + 13{\cdot}5X$$

where L is licence duty (£) (1973 tax level), and X is unladen weight (measured in units of 5 cwt or 0·25 tons), clearly does not reflect the cost pattern suggested by the AASHO study. A comparison of the 1973 excise licence duty paid and the new revised road wear and tear tax is shown in figures 2 and 3. Amongst articulated vehicles, for three- and four-axle units the new tax would be appreciably less than the excise licence duty for lighter vehicles, although more for the heaviest vehicles. For five-axle units the new tax would be significantly less. Amongst rigid vehicles, for two-axle units the new tax would be more favourable, and for three- and four-axle units only a greater burden at the top end of their weight range.

For most roads there is a large gap between total costs and marginal costs. Also, it is only infrequently that central or local government has to consider whether rising traffic on any particular road is going to necessitate long-run capital expenditure rather than the relatively small short-run marginal cost of maintaining a road. Thus Objective (i) as the basis of a charging system may have important 'efficiency' advantages.

The administrative costs of implementing the proposed tax are likely to be small relative to the direct and indirect benefits of the system. Evasion of existing excise licence duty by road users is estimated at between 1 and 2 per cent only, and it costs about £7 million to collect.[3] The need for some sort of registration exists independently of charges, for example to maintain a record of vehicle ownership, and for the police.

Thus the proposed new tax would result in at least one hidden cost being reflected by the tax structure. However, unlike fuel tax the new tax would be a lump sum tax levied annually and not directly related to the cost of a particular additional vehicle journey. Unfortunately as yet

## Table 32 A road wear and tear tax for commercial vehicles

| Vehicle category (GVW) (tons) | Estimated[a] standard axle laden ($A_{Li}$) | Estimated[a] standard axle unladen ($A_{ui}$) | No. of[a] vehicles ($V_i$) | Annual average kms laden[a] ($k_{Li}$) | Annual average kms unladen[a] ($k_{ui}$) | Estimated road wear and tear tax per vehicle, p.a. (£) |
|---|---|---|---|---|---|---|
| *Two-axle rigid* | | | | | | |
| 5 | 0·0162 | 0·0012 | 34,000 | 12,277·8 | 4,822·2 | 1·00 |
| 6 | 0·0651 | 0·0026 | 31,300 | 13,211·2 | 5,188·8 | 3·00 |
| 7 | 0·1279 | 0·0040 | 26,400 | 13,785·6 | 5,414·4 | 6·00 |
| 8 | 0·2212 | 0·0058 | 33,600 | 13,763·4 | 6,536·6 | 10·50 |
| 9 | 0·3898 | 0·0090 | 31,800 | 14,509·2 | 6,890·8 | 20·00 |
| 10 | 9·5645 | 0·0133 | 31,600 | 15,526·2 | 7,373·8 | 30·50 |
| 11 | 0·8500 | 0·0178 | 30,150 | 16,611·0 | 7,889·0 | 49·50 |
| 12 | 1·0638 | 0·0237 | 22,800 | 17,695·8 | 8,404·2 | 66·00 |
| 13 | 1·4532 | 0·0275 | 27,600 | 18,196·2 | 8,983·8 | 92·00 |
| 14 | 1·8104 | 0·0325 | 9,600 | 19,828·0 | 10,579·2 | 125·00 |
| 15 | 2·2371 | 0·0385 | 8,100 | 22,037·6 | 11,762·4 | 172·00 |
| 16 | 2·7050 | 0·0442 | 91,000 | 27,840·4 | 14,859·6 | 262·50 |
| *Three-axle rigid* | | | | | | |
| 15 | 0·4995 | 0·0102 | 1,000 | 17,734·4 | 9,465·6 | 31·00 |
| 16 | 0·6544 | 0·0120 | 450 | 19,951·2 | 10,648·8 | 45·50 |
| 17 | 0·7702 | 0·0155 | 1,400 | 20,994·4 | 11,205·6 | 56·50 |
| 18 | 1·0044 | 0·0201 | 1,000 | 22,559·2 | 12,040·8 | 79·00 |
| 19 | 1·2131 | 0·0239 | 350 | 24,645·6 | 13,154·4 | 104·50 |
| 20 | 1·5423 | 0·0283 | 14,400 | 26,014·8 | 13,885·2 | 140·00 |
| 21 | 1·9465 | 0·0336 | 400 | 31,491·6 | 16,808·4 | 213·50 |
| 22 | 2·4375 | 0·0398 | 10,800 | 32,404·4 | 17,295·6 | 275·50 |
| 23 | 2·8101 | 0·0463 | 100 | 33,269·0 | 18,231·0 | 326.00 |
| 24 | 3·4527 | 0·0517 | 2,100 | 34,431·8 | 18,868·2 | 414·00 |
| *Four-axle rigid* | | | | | | |
| 24 | 1·7112 | 0·0220 | 12,200 | 32,106·2 | 17,593·8 | 191·00 |
| 25 | 2·1578 | 0·0244 | 1,000 | 33,269·0 | 18,231·0 | 249·50 |
| 26 | 2·4188 | 0·0273 | 2,500 | 36,886·6 | 20,213·4 | 310·00 |
| 27 | 2·7078 | 0·0302 | 100 | 38,501·6 | 21,098·4 | 362·50 |
| 28 | 3·0260 | 0·0335 | 400 | 40,052·0 | 21,948·0 | 421·50 |
| 29 | 3·3770 | 0·0372 | 200 | 42,119·2 | 23,080·8 | 494·50 |
| 30 | 3·7630 | 0·0410 | 500 | 44,703·2 | 24,496·8 | 584·50 |

| | | | | | | |
|---|---|---|---|---|---|---|
| 10 | 0·0911 | 0·0087 | 5,000 | 6,520·0 | 3,480·0 | 2·00 |
| 11 | 0·1356 | 0·0106 | 650 | 7,302·4 | 3,897·6 | 3·50 |
| 12 | 0·1995 | 0·0128 | 900 | 8,932·4 | 4,767·6 | 6·50 |
| 13 | 0·2795 | 0·0150 | 500 | 11,018·8 | 5,881·2 | 11·00 |
| 14 | 0·3814 | 0·0175 | 500 | 13,628·8 | 7,273·2 | 18·50 |
| 15 | 0·5095 | 0·0205 | 500 | 16,300·0 | 8,700·0 | 29·50 |
| 16 | 0·6674 | 0·0241 | 4,700 | 20,994·4 | 11,205·6 | 49·50 |
| 17 | 0·8774 | 0·0271 | 600 | 23,080·8 | 12,319·2 | 71·00 |
| 18 | 1·1358 | 0·0312 | 3,600 | 25,167·2 | 13,432·8 | 100·00 |
| 19 | 1·4495 | 0·0359 | 2,600 | 25,688·8 | 13,711·2 | 130·50 |
| 20 | 1·8255 | 0·0411 | 4,900 | 26,210·4 | 13,989·6 | 167·50 |
| 21 | 2·2719 | 0·0470 | 100 | 26,210·4 | 13,989·6 | 208·00 |
| 22 | 2·7971 | 0·0540 | 1,700 | 26,732·0 | 14,268·0 | 261·00 |
| 23 | 3.4108 | 0·0612 | 500 | 27,318·8 | 14,581·2 | 325·00 |
| 24 | 4·1213 | 0·0689 | 9,100 | 27,384·0 | 14,616·0 | 393·50 |
| *Four-axle articulated* | | | | | | |
| 22 | 1·0826 | 0·0452 | 1,300 | 22,868·4 | 12,531·4 | 87·50 |
| 23 | 1·3493 | 0·0458 | 100 | 23,385·2 | 12,814·8 | 111·00 |
| 24 | 1·6692 | 0·0555 | 2,900 | 24,935·6 | 13,664·4 | 146·50 |
| 25 | 1·9229 | 0·0669 | 1,400 | 27,067·4 | 14,832·6 | 183·00 |
| 26 | 2·3414 | 0·0673 | 2,400 | 32,235·4 | 17,664·6 | 265·00 |
| 27 | 2·7978 | 0·0682 | 500 | 33,269·0 | 18,231·0 | 326·00 |
| 28 | 3·3446 | 0·0687 | 4,300 | 34,819·4 | 19,080·6 | 407·00 |
| 30 | 3·9370 | 0·0703 | 6,200 | 37,984·8 | 20,815·2 | 521·50 |
| 31 | 4·2871 | 0·0713 | 100 | 40,052·0 | 21,948·0 | 598·50 |
| 32 | 4·7268 | 0·0722 | 34,000 | 41,602·4 | 22,797·6 | 685·00 |
| *Five-axle articulated* | | | | | | |
| 32 | 2·2417 | 0·0337 | 3,600 | 40,568·8 | 22,231·2 | 317·00 |
| (1) | | | | | | |
| 32 | 2·2220 | 0·0697 | 800 | 40,568·8 | 22,231·2 | 314·50 |

1. Tri-axle Trailer

Source: [a]Derived from a survey of vehicle specifications, Department of the Environment, *The continuing road goods survey* (HMSO).

no measures of dispersion are available for the data derived from the *Continuing road goods survey* and used to calculate the new road 'wear and tear' tax in table 32. Inefficiency could result if the degree of dispersion for distance travelled by individual vehicles within each weight category is significant. It is disappointing to note that this gap in

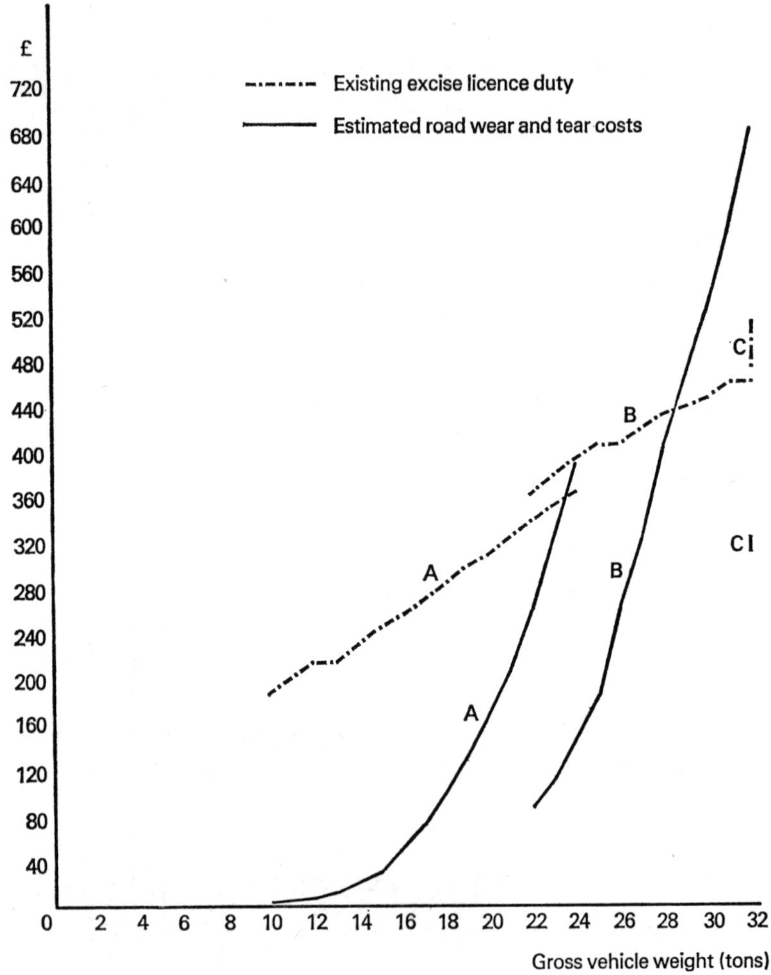


*Figure 2. Comparison of existing excise licence duty and estimated road wear and tear costs: articulated vehicles (£ per vehicle per annum). (A=three axles, B=four axles, C = five axles.)*

Ministry statistics, and the importance of having estimates of variance for kilometres travelled per vehicle to evaluate this form of tax was emphasized almost ten years ago.[11] Also there is a wide range of variation in the public costs of providing roads of different types. The maintenance cost per 1,000 vehicle miles for trunk roads is just over 10 per cent that of Class III roads (see table 33). It could be argued that a charging system should encourage users to travel where marginal wear and tear costs are low and *vice versa*. The proposed road 'wear and tear' tax would not of course do this, and inevitably efficiency losses would arise.

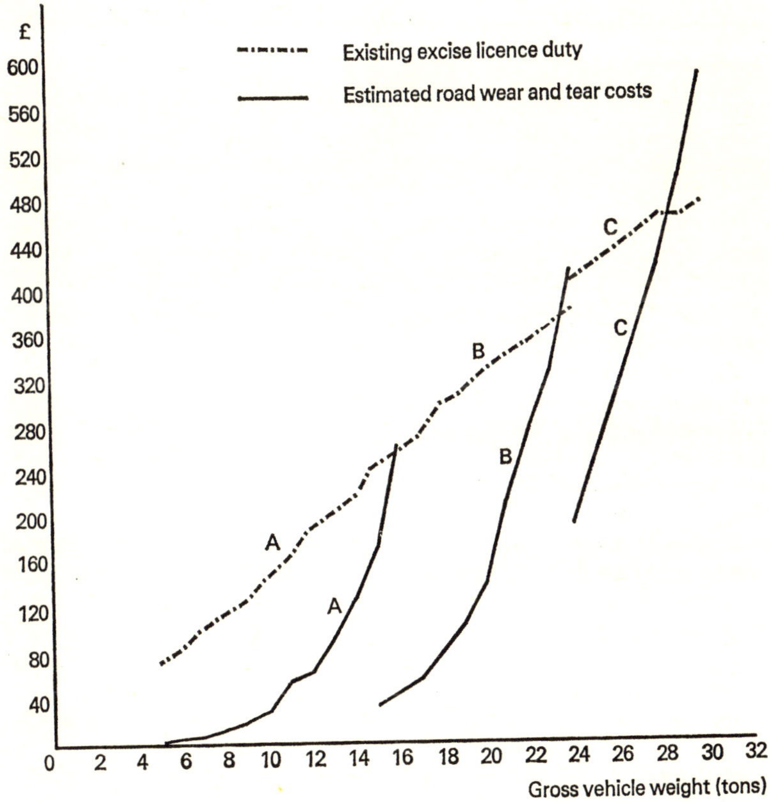


*Figure 3. Comparison of existing excise duty and estimated road wear and tear costs: rigid vehicles (£ per vehicle per annum). (A = two axles, B = three axles, C = four axles.)*

A related problem is the flexibility of a road 'wear and tear' tax over a period of time as conditions change. As A. A. Walters has observed, 'The necessity for political approval for any change in the taxation system means that the price is quite rigid for long periods. While the taxation and road pricing system is controlled directly by the legislature this long lag seems to be part of the nature of the political process. A consensus of opinion requires a long time to organise and secure.'[12]

**Table 33 Maintenance costs by type of road**

| | Cost per mile of road (£). | Cost per 1,000 vehicle miles (p). |
|---|---|---|
| Trunk roads | 1,047 | 25 |
| Classified roads | | |
| I | 883 | 42 |
| II | 670 | 48 |
| III | 394 | 212 |

Source: *The Road Track Cost Report* (HMSO, 1968).

An evaluation of the efficiency of the proposed new tax should be based on accurate knowledge of elasticities of substitution. From the point of view of efficiency, if there is no elasticity of substitution it does not matter what charges are made. If elasticities are high it matters correspondingly more that relative prices should be 'right'. Between goods carried in heavy or light commercial vehicles elasticities are likely to be high. In contrast, empirical evidence indicates that price elasticities of demand for road haulage service are low. A U.K. study of consignments shipped in the manufacturing sector revealed price not to be an important factor influencing the distribution of traffic between modes.[13] Price substitution elasticities were −0·21 for own-account versus professional operations, and a similar result was found for professional road haulier versus rail, although this was not significant at the 5 per cent level. Low elasticities may also be implied from the study's result that only for about 25 per cent of consignments dispatched was the shipper able to give *any* indication of the charge by an alternative mode.

Given implementation of the proposed marginal road 'wear and tear' cost pricing structure (and assuming that the objective is to maximize social surplus) competition would achieve an efficient allocation of resources within the transport sector. If marginal cost pricing were to exist throughout the economy this would result in a Pareto-optimal allocation of resources, such that no further change in the allocation of resources could make someone better off without making someone else worse off. Marginal cost pricing is not, however, applied consistently either across the economy as a whole or in other transport sectors. Analysis of second-best effects throughout the economy is very complicated.[14] Empirical investigations are few, and none have taken place in transport. Nevertheless, some indications of the likely implications of adopting marginal cost pricing along the lines proposed may be ventured.

The practical question to be determined is whether or not there is *prima facie* evidence for thinking that there is a relationship of substitution or complementarity such that net welfare losses elsewhere will more than offset the welfare gain from implementing the marginal cost pricing proposals. Since inland waterways are as yet relatively unimportant in the U.K., the pertinent relationship is that between price and marginal cost on rail. Implementing a road 'wear and tear' tax will be an increase in efficiency provided that price does not exceed marginal cost on the railways. Since this condition is not fulfilled for rail freight charges, there is a high probability that reducing road charges towards marginal road wear and tear social cost, would result in an overall loss in efficiency. However, given that road wear and tear understates the true marginal cost, which should also include environmental cost, congestion, and a shadow price for foreign exchange spent on fuel imports, then we may draw a more optimistic conclusion about the effect on efficiency.

A key factor is the marginal rate of substitution between road and rail, which we have noted appears to be low. On most roads and for most classes of traffic a change in road prices is unlikely to lead to an appreciable shift of traffic between road and rail. This factor has led Foster, in his reassessment of the transport problem, to shift his priorities.[15]

> As the amount of road traffic declines for which the railways are an effective competitor, so the probability declines that introducing marginal social cost pricing on roads, without similarly introducing marginal social cost pricing on rail, will result in an overall welfare loss.
>
> . . . with hindsight, even then, (1963), I suspect I was giving the traditional policy of coordination of criteria in transport more weight then its effect on the efficient allocation of resources deserved.

The problem of co-ordinating road and rail pricing should not therefore be such a serious obstacle to progress in implementing a limited form of road infrastructure pricing, although where road competes directly with rail freight operations, the same pricing rules should be adopted.

A potentially more serious weakness of the proposed new tax is that in calculating the road wear and tear factor the average standard axles are based on 'nominal loads' derived from a survey of vehicle specifications. Given that the objective of the new tax structure is to charge a price reflecting the actual road wear and tear imposed by a vehicle category, then the measured 'on the road' or 'effective load' basis for estimating standard axles has obvious advantages.

Unfortunately, data available on 'effective loads' is very limited. In table 34 a comparison is made between average standard axles based on nominal loads as in table 32, and 'effective loads' derived from a limited sample of axle loads as measured for different vehicles on the road at test sites on the M1, A40, and A30.[16]

Using the 'effective load' basis for estimating standard axles would result in a much more skewed distribution of the road wear and tear costs. Examining vehicles actually operating on the roads would indicate that the heavy vehicles have a much higher damage factor and therefore should pay much more than indicated by the theoretical approach. However, the sample was small, and a far bigger survey covering different types of road would be required before drawing up a system of charges based on the effective load approach. Also the sample, since it was based on actual loads, includes overloaded vehicles breaking the law. The effective load of legally operated vehicles is more likely to be closer to the ratio of the nominal loads.

An interesting conclusion from the work reported by Croney and Watkins at one of the test sites (the A1 at Alconbury Hill) is that less than 4 per cent of the dynamic axle loads measured are in excess of the legal limit, but these will cause about 30 per cent of the damage to the road. In 1972 the number of people fined for exceeding the maximum gross weight, or axle weight, was 4,027. The total amount of fines imposed was £49,002, making an average fine per head of £12, and there were 30 endorsements without disqualification.[17] If, strictly for fun, we divide up 30 per cent of the road maintenance expenditure, fines would not be £12 per head, but £17,000! This policy would be consistent with the principle that a price should be charged to the user equal to the demand he makes on society's resources – unfortunately with such a deterrent system, society would probably have to provide more resources in the form of debtors' prisons. Croney and Watkins compare the percentage of axles overweight with the percentage of vehicles overweight at several test sites, and conclude that a significant proportion of overweight axles are due to unsatisfactory load distribution on the vehicle rather than excessive gross load. There is also the problem of those 'guilty' but not caught, although with a few exemplary fines which

**Table 34 The road wear and tear tax using 'nominal load', compared with 'effective load' basis for measuring standard axles**

| Vehicle category | Estimated average standard axles using nominal loads | Ratio of average standard axles using nominal loads | Estimated average standard axles using effective loads | Ratio of average standard axles using effective loads |
|---|---|---|---|---|
| Two-axle rigid | 1·12 | 1·00 | 0·39 | 1·00 |
| Three-axle rigid | 1·90 | 1·71 | 0·86 | 2·21 |
| Four-axle rigid | 1·98 | 1·78 | 1·90 | 4·87 |
| Three-axle artic. | 1·64 | 1·47 | 0·98 | 2·51 |
| Four-axle artic. | 4·12 | 3·69 | 2·73 | 7·00 |
| Five-axle artic. | 2·28 | 2·04 | 2·89 | 7·41 |

Source: D. Croney, and L. H. Watkins, 'Heavy vehicles – research aspects', Symposium on the Heavy Vehicle, Institute of Highway Engineers, 1974.

**Table 35 Revenue/cost ratios for commercial vehicles, 1972–3**

| Vehicle category and weight (GVW) | Road maintenance cost | Capital cost: Gross tonne kms | Capital cost: PCU vehicle kms | Total cost | Fuel tax revenue | Excise licence duty | Total revenue | Revenue: cost |
|---|---|---|---|---|---|---|---|---|
| *Two-axle rigid* | | | | | | | | |
| 5 tons | 1 | 26 | 87 | 114 | 141 | 78 | 219 | 1·9:1 |
| 16 tons | 263 | 214 | 218 | 695 | 543 | 257 | 800 | 1·2:1 |
| *Three-axle rigid* | | | | | | | | |
| 16 tons | 45 | 153 | 156 | 355 | 389 | 257 | 646 | 1·8:1 |
| 24 tons | 414 | 393 | 272 | 1079 | 828 | 378 | 1206 | 1·1:1 |
| *Four-axle rigid* | | | | | | | | |
| 24 tons | 191 | 370 | 253 | 814 | 772 | 405 | 1177 | 1·3:1 |
| 30 tons | 585 | 637 | 353 | 1575 | 1209 | 473 | 1683 | 1·1:1 |
| *Three-axle artic.* | | | | | | | | |
| 16 tons | 49 | 162 | 164 | 375 | 409 | 257 | 666 | 1·8:1 |
| 24 tons | 393 | 310 | 214 | 917 | 652 | 365 | 1017 | 1·1:1 |
| *Four-axle artic.* | | | | | | | | |
| 24 tons | 146 | 286 | 197 | 629 | 600 | 392 | 992 | 1·6:1 |
| 32 tons | 685 | 623 | 328 | 1636 | 1286 | 459 | 1745 | 1·1:1 |
| *Five-axle artic.* | | | | | | | | |
| 32 tons | 317 | 616 | 320 | 1253 | 1254 | 473 | 1727 | 1·4:1 |
| 32 tons (with tri-axle trailer) | 315 | 610 | 320 | 1245 | 1254 | 513 | 1767 | 1·4:1 |

reflected the true costs to the community this number is likely to decline drastically. These findings indicate the need for urgent action since more efficient loading and vehicle design would reduce road maintenance costs significantly. Also there could be a high rate of return on resources allocated to more effective enforcement of existing legislation. Assuming the excise licence duty is revised along the lines proposed and related more closely to road wear and tear, then all commercial vehicle operators, as well as the community as a whole, could benefit.

*Objective (ii)*
Devising a charging system to reflect capital costs as well as short-run maintenance costs encounters many problems. Implementation of a long-run marginal cost pricing system would include only those costs that would have to be incurred to produce an additional unit of output on an entirely new infrastructure. Given the problems of measuring the replacement cost of roads, the *Road Track Cost Report* treats capital expenditure on roads as current expenditure, to be recovered in the year in which it was incurred. This method of treating capital costs also has its weaknesses,[18] but it retains the advantage of the LRMC charging principle of reflecting actual commitment of resources, and disregarding past expenditures.

Having decided on the current expenditure method of treating capital costs the next problem is to determine the capital costs attributable to commercial vehicle categories under consideration. The British Railways Board, in its evidence to the Geddes Committee, argued that over 70 per cent of the cost of building a high-capacity road would be saved if it did not need to be designed for heavy vehicles.[19] This estimate received no support, and further work by the Ministry of Transport suggested that 17½ per cent was a more realistic potential saving.[20] On this basis, using 1972–3 estimates, £94·1 million may be allocated to commercial vehicles. This proportion of capital expenditure on construction and major improvements may then be attributed between individual categories on the basis of gross-tonne-kilometres, as in table 35. A residual of capital costs amounting to 82½ per cent is then allocated amongst all vehicles on the basis of PCU-kilometres. The latter are generally accepted as a measure of the space occupied by, and the effect on traffic flow of, different types of vehicle.
Thus capital cost per vehicle in category i is =

$$C_g \left[\frac{W_{Li}K_{Li} + W_{ui}K_{ui}}{\sum_{i=1}^{56} (W_{Li}K_{Li}V_i + W_{ui}K_{ui}V_i)}\right] + C_R \left[\frac{P_{Li} + P_{ui}}{\sum_{i=1}^{56} (P_{Li}V_i + P_{ui}V_i)}\right]$$

where $C_g$ = Capital expenditure directly attributable to commercial vehicles

$C_R$ = Residual of capital expenditure attributable to commercial vehicles

$W_{Li}$ = Gross Vehicle Weight of category i

$K_{Li}$ = Annual average kilometres laden

$K_{ui}$ = Annual average kilometres unladen

$P_{Li}$ = PCU kilometres laden

$P_{ui}$ = PCU kilometres unladen

$V_i$ = Number of vehicles in each of the 56 gross vehicle weight categories distinguished

As might be expected from both weighting systems, the burden of capital expenditure increases with vehicle weight. In fact, as shown in table 34 the capital charge for the heaviest rigid and articulated vehicle is just under and just over ten-fold, respectively, the capital charges for the lightest vehicle.

**Table 36 Passenger car unit (PCU) weighting system**

| | Rural | Urban |
|---|---|---|
| Private car | 1 | 1 |
| Motorcycle combination, solo, moped, taxis, light goods vehicles less than 30 cwt ULW | 1 | ·75 |
| Heavy goods vehicles more than 30 cwt ULW | 3 | 2 |
| Bus, coach | 3 | 3 |
| Pedal cycle | ·5 | ·33 |

By combining the different components of costs attributable to commercial vehicle categories estimated in the previous sections, and relating them to their fuel tax revenue, and excise duty, it is possible to give more recent estimates of revenue/cost ratios. The results shown in tables 32–5 are similar to those of 1965–6 and 1971 in that for each category revenue exceeds costs. Lighter vehicles pay proportionately more in revenue than costs allocated to them, while, for heavy goods vehicles, the situation is reversed. From the point of view of equity, at least, every category of commercial vehicle is paying its way, although lighter vehicle users might argue that it is inequitable that their excess revenue over cost should be greater than heavier vehicles, and these differentials could distort optimal fleet choice. Even accepting the weaknesses of the very arbitrary current expenditure method as a proxy for LRMC, this is an appropriate basis for pricing policy only if it is believed either that capacity is optimal, or that it will be optimally adjusted within a sufficiently short space of time to make the expense of charging short-run marginal cost in the interim, and then changing price again, not worthwhile. In the optimal position LRMC and SRMC will always be equal, so that if LRMC is constant with respect to traffic flow, and therefore equals long-run average cost, the tax paid by road users will exactly equal the cost of providing and maintaining the road.

The reality of road expenditure, which is affected by the interaction of pressure groups, and which is characterized by joint costs and indivisibilities, undermines the contribution of theory as a guide to policy. In the case of indivisibilities, Williamson has shown that it is still possible to determine the optimal level of capacity using LRMC, but that generally a surplus or deficit will result.[21] Pricing and investment policy in respect of roads is further complicated because demand in one place cannot be met by facilities located in another.

*Objective (iii)*
From the foregoing analysis there is no reason to suppose that implementing optimal road taxes will exactly cover the costs of providing and maintaining roads. In fact large surpluses may arise, particularly in urban areas, or large deficits, particularly in rural areas, and there is no reason why one should offset the other. Given that a congestion element is not introduced into the optimal tax, the likely outcome is a deficit and this may have undesirable effects. It may be considered inequitable to put all or any of the burden of infrastructure upon taxpayers as a whole. Incorrect investments in infrastructure may be made as a result of political pressures that may be brought to bear where public expenditure is involved. Subsidies intended to finance the deficit of infrastructure may also be used to cover inefficient operation, and may enable inefficiency to continue. Finally, competition may be distorted if some modes of transport are relieved of the financial burdens connected with their infrastructures, whilst other modes have to carry them.

Moreover, as we have already noted, an objective of vehicle taxation may be to raise revenue for the Exchequer, as well as attempting to allocate resources more efficiently. If the existing excise licence duty was revised to accord with Objective (i) to reflect short-run marginal road wear and tear cost then there would be a considerable reduction in revenue collected. The Exchequer is, however, highly unlikely to accept a reduction in revenue.

An analysis of efficient prices which would maximize social surplus, subject to an additional budget constraint, concludes that there are a wide variety of output combinations and sets of prices that can be chosen, each of which would just meet the net revenue constraint.[22] However, assuming the net revenue constraint, there is one efficient solution to the problem of which of these gives the best use of resources to serve consumer preferences: 'the surprisingly simple optimality rule asserts that Pareto optimal utilisation of resources in the presence of an absolute profit constraint requires (considering the substitution effects alone) that all outputs be reduced by the same proportion from the quantities that would be demanded at prices equal to the corresponding marginal costs.'

The implication of this pricing strategy is that output for which demand is more elastic should have prices approximating to marginal cost more nearly than those for which demand is more inelastic. For example, road freight on routes where elasticity of substitution was low should be charged a relatively higher price than traffic where inland waterway or rail was a closer substitute. If a net revenue constraint were to be laid down it would be necessary to determine the time period for recovery, and the pattern of cost recovery over the period. If a financial balance constraint was imposed, a further problem would be to determine the extent of the regional or modal sub-network over which the constraint should operate. A road improvement may both lower receipts and by causing an increase in maintained road length raise costs. In the absence of a large generation of new traffic this would necessitate an increase in unit price throughout the sub-sector.[23] Exactly how this would affect different types of traffic or different routes would depend on how it was decided to recover the extra revenue required. Clearly it would require a far more complex charging system than the revision of vehicle excise licence duty as proposed in this Chapter.

It is the policy of the EEC that all its members should eventually harmonize their methods of charging for state-provided infrastructure. A tax based on marginal cost similar to that discussed in this Chapter has been proposed for road users. But the EEC road use tax would replace both vehicle licence duty and fuel tax. This could result in lowering the tax burden on vehicles unless charges for social costs (other than road wear) were made. The EEC also favours the use of surplus tax revenue for the improvement of infrastructure, a procedure which would conflict with the dislike of the British Treasury for the 'hypothecation' of tax revenue (i.e., its commitment to some specified form of expenditure).

## Summary

*1. The problem of defining and efficiently and equitably allocating track costs has occupied a central position in transport economics. As an important element in total marginal social cost along with other community costs such as noise, smoke, visual intrusion, etc., track costs are important in an examination of transport and the environment. Given indivisibilities, joint costs, and the free-rider problem, public intervention is essential. However, even with perfect knowledge, we still face a choice between direct regulation by the Government, or indirect manipulation of a vehicle operator through setting fees, or taxes, to reflect accurately road wear and tear caused by his vehicle.*

*2. A road wear and tear tax for commercial vehicles, to replace existing excise licence duty, is derived from the estimated number of vehicles in each group, kilometres travelled laden and unladen, and number of average standard axles related to the fourth power law. The tax bears much more heavily on the heavier vehicles. A 5-ton gross vehicle weight two-axle rigid would pay only £1, as against £685 for a 32-ton gross weight four-axle articulated vehicle. The number of axles is crucial: a 24-ton four-axle articulated unit would pay £147, whereas a 24-ton three-axle rigid would pay £414.*

*3. The advantages and disadvantages of the road wear and tear tax for commercial vehicles are examined. Such a tax would reflect road wear and tear costs far more accurately than existing taxes; given the large gap between total costs and marginal costs for roads, the short-run marginal cost basis of tax should increase efficiency in resource allocation; administrative costs and evasion are not likely to be significant. Disadvantages of the tax include its losses in efficiency because of averaging within vehicle groups, and types of roads; its annual lump sum payment basis; evidence of low substitution elasticities; distortions arising from different pricing and investment policies on other modes.*

*4. Available evidence indicates that a small number of overloaded commercial vehicles cause a disproportionately large share of road wear and tear costs. If the fines of those prosecuted were made to reflect the true cost imposed on society, average fines would rise from £12 to £17,000. Tests indicate that a significant proportion of over-weight axles are due to bad load distribution rather than excessive gross load. Provision of information to operators on loading techniques could help reduce road wear and tear costs imposed on society, as also would more resources allocated to policing existing legislation.*

*5. An attempt is made to allocate capital expenditure on road construction and major improvement, arising from each commercial vehicle category. The Ministry of Transport estimates that 17½ per cent of capital expenditure may be attributed to commercial vehicles, which is allocated on the basis of gross-tonne-kilometres, and the residual capital costs allocated amongst all vehicles on the basis of PCU-kilometres. The capital charge for the heaviest rigid and articulated vehicle is just under, and just over, ten-fold respectively the capital charges for the lightest vehicles. Even accepting the weaknesses of current expenditure as a proxy for long-run marginal cost, this is an appropriate basis for pricing policy only if capacity is optimal, or if it will be optimally adjusted within a sufficiently short space of time to make the expense of charging short-run marginal cost in the interim, and then changing price again, not worth while.*

*6. There is no reason why introducing optimal road taxes will exactly cover the cost of providing and maintaining roads, and surpluses or deficits may arise. To avoid the disadvantages of such imbalances, and perhaps also to meet a fiscal requirement set by the Exchequer, an additional net revenue constraint may be imposed. Recent theoretical advances indicate that the most efficient charging system may then arise from charging 'what the market will bear'.*

Chapter six

# Accidents

## 1 Introduction

Some people would make the value judgment that accidents constitute the most serious of all the social costs of transportation. While many more people are affected by noise each day than would be involved in accidents during a year, the risk to life and limb is obviously a disbenefit of a different order of magnitude from that caused by noise or visual intrusion. Those who object to the use of economic measurements for evaluating proposed environmental improvements sometimes feel particularly strongly about the propriety of putting money values on human life and suffering. But, as has already been argued in Chapter One, choices do have to be made between reducing death and injury rates and obtaining other forms of satisfaction. Society does not regard life as being of infinite value. Accident rates would undoubtedly be reduced if the speed limit on all roads was reduced to 20 mph, but it seems reasonably certain that a referendum on a proposal to lower maximum permitted speeds to this level would yield a majority of 'No' answers. Since the benefits of expenditure on reducing accident rates has to be compared with the benefits from alternative forms of public expenditure it is necessary to express the costs of fatal and injury accidents in the common currency of money. The issue of whether the money values used in Britain to represent the costs of accidents are adequate will be discussed at the end of this chapter.

A full study of transportation accidents would be a major enterprise and it is necessary to be fairly selective in this account of the problem. The main omission is any attempt to evaluate detailed methods of improving accident rates or reducing the severity of accidents. The discussion is centred mainly on the more important policy issues. As elsewhere in this book the problems of lorries, which receive much popular blame for accident involvement, are used to illustrate the issues.

## 2 The accident involvement of different classes of road vehicle

The figures in table 37 show all road death and injury accidents in 1974 classified according to the class of road user involved. These figures give some idea of the magnitude of the problem, with nearly 7,000 people killed and over 80,000 seriously injured on the roads in one year. It is not easy to discover the extent to which accident rates can be controlled. There may be a minimum level of accident occurrence which must be accepted as virtually unavoidable. But variations in accident rates over time, and between different countries, do suggest that accident prevention policies can be effective. Long-term trends in accident rates need to be interpreted cautiously since there have been considerable fluctuations in the volume of traffic. However, a comparison of accident rates in 1934 and 1973 does suggest very strongly that there has been considerable success in reducing the likelihood of road accidents. In 1934, when the population of Great Britain was 45·4 million and there were 2·4 million vehicles with current licences, 7·3 thousand people were killed on the roads and total casualties numbered 239,000. The equivalent figures in 1973, when the population was estimated to be 54·4 million and 16.6 million vehicles were licensed were 7·4 thousand deaths and 353,000 total casualties. Total casualties increased by 99·4 per cent between 1949 and 1973 while motor traffic vehicle mileage increased by 525 per cent.[2] Some of the variation in accident rates between different countries is illustrated by the figures in table 38.

The figures have been adjusted to allow for different practices in defining a road fatality. The road deaths per 10,000 vehicles figure does not allow for differences in annual vehicle mileages. It is impossible to say how far these different accident rates reflect different policies towards accident prevention. Clearly driving habits, which may not be influenced by policy decisions, are an important factor influencing accident rates. Nevertheless it seems very likely that accident prevention policies are an important factor in explaining these inter-temporal and international differences. If the British annual death rate per 10,000 vehicles was the same as that in West Germany (in 1972) then, in 1974, 16,600 people would have been killed on British roads. If there is a likelihood that at least some of the 9,200 people who are still alive had their lives 'saved' as a result of accident prevention policies then it is obviously very well worth while to find out whether policies can be adopted to reduce accident rates to even lower levels.

**Table 37 Road injury accidents by severity and class of road user, 1974**

| Class of road user | Killed | Seriously injured | Slightly injured | Total |
|---|---|---|---|---|
| Pedestrians | 2,648 | 21,051 | 50,909 | 74,608 |
| Pedal cyclists | 282 | 4,167 | 14,453 | 18,902 |
| Users of mopeds, motor scooters and motorcycles | 798 | 13,919 | 32,154 | 46,871 |
| Users of cars and taxis | 2,709 | 36,565 | 111,078 | 150,352 |
| Users of goods vehicles | 334 | 4,427 | 13,575 | 18,336 |
| Users of public service vehicles | 69 | 1,549 | 12,365 | 13,983 |
| Users of other vehicles | 46 | 461 | 1,409 | 1,916 |
| Total | 6,886 | 82,139 | 235,943 | 324,968 |

Source: *Road Casualties in Great Britain, Year 1974* (Department of the Environment, 1975).

**Table 38 International comparisons of accident rates, 1972**

| Country | Number of road deaths | Road deaths per 10,000 vehicles | Car user deaths per 100 million car miles |
|---|---|---|---|
| Sweden | 1,194 | 4.0 | — |
| Great Britain | 7,763 | 4·8 | 2·7 |
| U.S.A. | 56,300 | 5·0 | — |
| France | 16,545 | 8·0 | 10·0 |
| West Germany | 18,811 | 10·0 | 6·0 |
| Spain | 4,453 | 11·0 | 14·0 |
| Portugal | 1,695 | 26·0 | 12·0 |
| Yugoslavia | 4,460 | 34·0 | 15·0 |

Source: *Road Accidents Great Britain, 1973* (Department of the Environment, Scottish Development Department, Welsh Office, 1974).

The figures in table 39 show the involvement of different types of vehicle in road accidents in 1973. These figures are greater than the number of injuries or deaths since in some cases more than one vehicle was involved in an accident. Thus 10,043 motor vehicles were involved in accidents in 1973 in which 6,771 people were killed. Vehicles which collided after the initial impact, causing injury, were not included unless they aggravated the degree or amount of injury. There is a slight discrepancy between the figures in tables 17 and 43 in the official 1973 accident statistics.[2] A better idea of the relative accident involvement of different types of vehicle can be obtained by relating accidents to road use. In table 40 injury accident involvement is related to annual vehicle mileage.

**Table 39 Vehicles involved in injury accidents, 1973**

| | Vehicles involved in | | | | |
|---|---|---|---|---|---|
| Type of vehicle | Fatal accidents Number | Serious accidents Number | Slight accidents Number | All accidents Number | % |
| Mopeds | 93 | 2,245 | 5,876 | 8,214 | 2·1 |
| Motor scooters | 80 | 1,785 | 4,306 | 6,171 | 1·6 |
| Motorcycles | 713 | 10,734 | 20,570 | 32,017 | 8·1 |
| Cars and taxis | 6,414 | 71,162 | 191,977 | 269,553 | 67·9 |
| Public service vehicles | 418 | 3,234 | 14,091 | 17,743 | 4·5 |
| Goods vehicles not over 1·5 tons ULW | 762 | 8,355 | 23,152 | 32,269 | 8·1 |
| Goods vehicles over 1·5 tons ULW | 1,384 | 6,872 | 14,886 | 23,142 | 5·8 |
| Other vehicles | 179 | 1,823 | 5,544 | 7,546 | 1·9 |
| All motor vehicles | 10,043 | 106,210 | 280,402 | 396,655 | 100·0 |

Source: *Road Accidents Great Britain, 1973* (Department of the Environment, Scottish Development Department, Welsh Office, 1974).

**Table 40 Vehicles involved in injury accidents per hundred million miles travelled, 1973**

| Vehicle type | All accidents | Fatal accidents | Serious accidents |
|---|---|---|---|
| Goods vehicles over 1·5 tons ULW | 180 | 11 | 54 |
| Cars and taxis | 224 | 5 | 59 |
| Goods vehicles not over 1·5 tons ULW | 244 | 6 | 63 |
| Public service vehicles | 797 | 19 | 145 |
| Pedal cycles | 942 | 16 | 216 |
| Motor scooters | 1,466 | 19 | 424 |
| Mopeds | 1,703 | 19 | 465 |
| Motorcycles | 2,285 | 51 | 766 |

Source: *Road Accidents Great Britain, 1973* (Department of the Environment, Scottish Development Department, Welsh Office, 1974).

The relative position of different types of vehicle has not changed greatly over the last ten years. The ranking according to 'all accident' involvement was very similar in 1963, except that public service vehicles had a higher involvement rate than pedal cycles and the relative positions of motor scooters and motorcycles in the ranking were reversed. The trend of accident involvement per hundred million miles travelled has been fairly consistently downwards for cars, both weight categories of goods vehicle, and (though less markedly) for public service vehicles. The trend for motorcycles, mopeds, and pedal cycles has fluctuated, but has been mainly upward.

Before discussing the significance of the figures in tables 39 and 40 some comment on the meaning of the word 'involvement', in relation to accidents, is necessary. There are two different types of 'involvement' which are sometimes confused both in discussions of accident problems and in the statistics. Some road accidents can be said to be caused by vehicles in the sense that the vehicle or its driver is directly responsible for the accident. Examples would be where a lorry's brakes fail on a hill and it crashes into parked cars, or where a car driver overtakes on a blind bend and causes a collision. The other type of involvement occurs when road users come into conflict, but when at least one of the vehicles cannot be said to have any responsibility for the accident other than that of being on the road. The other car hit by the overtaking motorist would clearly be involved but not responsible. Many accidents occur when vehicles hit other stationary vehicles which are legitimately parked. If a motorist loses control of his car and hits a lamp standard then the accident can be said to involve both the motorist and the lamp standard. The difference between the two types of involvement would be apparent in investigating possible measures for avoiding future motorist/lamp standard collisions when it would seem more reasonable to call for an improvement in driving techniques rather than the removal of all street lamps (although collapsible lamp standards have been developed).

The figures in table 40 indicate that motorcycles had by far the greatest accident involvement rate of any road vehicle type and that this applied to fatal, serious, and slight injury accidents. Public service vehicles also had relatively high involvement rates. 'Heavy' goods vehicles had the lowest overall involvement rate, but their rate was higher than that of cars and 'light' goods vehicles for fatal accidents. There are several possible explanations of these widely different accident involvement rates. One important distinction which is not shown by the figures is that between injury accidents to drivers or passengers of the vehicle type involved, and those caused to other people. In the case of pedal cycles and two-wheeled motor vehicles most of the injuries are suffered by the drivers themselves and one reason for the high injury accident rate is that drivers are relatively unprotected. An accident in which a car or lorry driver might be uninjured could result in serious injury for a motorcyclist. Although motorcycles and buses both have a relatively high accident involvement rate, the position of their drivers is quite different. In 1973 there were 642 serious injury casualties affecting motorcycle riders (per hundred million miles travelled) but only 5 affecting drivers of public service vehicles.[2]

A second factor which may explain the differential rates is the type of driver. Motorcycles tend to be driven by young men in the most accident-prone age groups. A third factor of importance is the type of road on which vehicles operate. Accident rates are obviously likely to be higher in built-up areas, where there are many pedestrians and mixed traffic, than they are on rural roads (although higher speeds on rural roads may lead to more serious accidents). The figures for accident involvement and vehicle mileage for built-up and non-built-up areas in 1973 were

| | |
|---|---|
| Built-up areas: | |
| Vehicles involved in all accidents | 293,151 |
| Estimated vehicle mileage | 74,669 (millions) |
| Non-built-up areas: | |
| Vehicles involved in all accidents | 103,504 |
| Estimated vehicle mileage | 76,552 (millions) |

Vehicles were thus nearly three times as likely to be involved in accidents in built-up areas (3·93 per million miles) than in non-built-up areas (1·35 per million miles). In non-built-up areas some roads are safer than others. Motorways with their absence of intersections have particularly low accident rates. In 1973 motorways accounted for about 7 per cent of all traffic but only 4·5 per cent of deaths and 2·2 per cent of serious injuries to permitted users of motorways.[2] This road-type factor must account for some of the differences in vehicle involvement rates. A heavy lorry operating mainly on motorways or major trunk roads is clearly likely to have a lower accident involvement rate than a stage bus running entirely on all-purpose roads in built-up areas.

A final possible factor might be vehicle size. During the *Living with the Lorry* study if was found that many people believed that lorries were particularly dangerous because of their size. Thus the Civic Trust report on heavy lorries argued that 'there is a strong case for suggesting that freight traffic should only be allowed to travel by heavy goods vehicle when it cannot be conveyed by an alternative transport mode that is less prodigal of human life and limb.'[3] The statistics which have been produced so far indicate that heavy lorries have a low overall accident involvement rate, but that their involvement in fatal accidents is rather higher. This issue is discussed further below with the aid of some more specific figures relating to accidents affecting cars. But first the problem of pedestrian accidents is considered in more detail.

The figures in table 41 show the vehicles involved in accidents between single vehicles and pedestrians in the years 1960 and 1967.

**Table 41 Pedestrians killed in single-vehicle accidents per hundred million miles travelled, by the class of vehicle involved**

| | Fatal pedestrian accidents per 100 million vehicle miles | |
|---|---|---|
| Vehicle involved | 1960 | 1967 |
| Car or taxi | 1·9 | 1·2 |
| Public service vehicle | 4·7 | 3·9 |
| Motorcycle | 4·2 | 4·5 |
| Motor scooter | 2·8 | 2·5 |
| Moped | 0·6 | 0·6 |
| Pedal cycle | 0·3 | 0·1 |
| Goods vehicle | 2·1 | 1·7 |

Source: H. D. Johnson and F. Garwood, *Notes on road accident statistics* (RRL Report LR394).

In relation to distance travelled, motorcycles, buses, and motor scooters were the vehicles most likely to be involved in a fatal accident with a pedestrian. As already argued, the 'road factor' must be taken into consideration in interpreting these figures: buses run a much higher proportion of their mileage on high-accident-rate urban roads than do heavy lorries. Since there are more goods vehicles than buses or motorcycles on the roads, the absolute number of pedestrians killed in lorry/pedestrian accidents is considerably higher than in motorbike/pedestrian or public service vehicle/pedestrian accidents. An analysis of 2,718 fatal pedestrian accidents in 1968 (total fatal pedestrian accidents in that year amounted to 2,762) showed that 164 pedestrians were killed in motorcycle/pedestrian accidents; 171 in public service vehicle/pedestrian accidents; 263 in light goods vehicle/pedestrian accidents; and 289 in medium and heavy goods vehicle/pedestrian accidents.[5] Measured by absolute numbers, however, the car was the most dangerous road vehicle for pedestrians, with 1,699 fatal car/pedestrian accidents in 1968. (58 pedestrians were also killed in accidents involving two cars.) The pedestrian fatalities in this 1968 data related to vehicle distances travelled is shown in table 42.

As might be expected the chances of being killed rather than seriously injured are greater for pedestrian accidents involving heavier vehicles.

Thus 1 in 30 pedestrian/vehicle accidents (leading to death or serious injuries) were fatal if the vehicle was a pedal cycle or moped; 1 in 10 if it was a car; and 1 in 5 if it was a heavy goods vehicle or a bus.[5]

**Table 42 Pedestrian fatalities in relation to distance travelled, 1968**

| Type of vehicle involved in accident | Estimated annual vehicle kms ($10^9$) | Pedestrian fatalities per $10^9$ vehicle kms |
|---|---|---|
| Pedal cycle | 5·0 | 3 |
| Moped | 0·8 | 4 |
| Other motorcycle | 3·9 | 42 |
| Car | 145·5 | 12 |
| Public service vehicle | 3·8 | 45 |
| Light goods vehicle | 17·7 | 15 |
| Other goods vehicle | 18·8 | 17·5 |

Source: *Contributions to the accident situation by all types of vehicle* (TRRL Leaflet LF 320; reproduced in *Towards safer road vehicles*, TRRL Report LR 481).

In so far as statistical evidence can be used to analyse the situation it would seem that the main reason why lorries may be a traffic hazard is not because they are poorly driven or badly maintained, but because they are large. It is indeed not necessary to find statistical evidence to demonstrate that in any collision between a lighter and heavier vehicle, the lighter one will tend to suffer most damage, or that the seriousness of an impact accident will tend to increase with the size and weight of the object that is hit. Speed and other variables being the same, the likelihood that the driver of a family saloon car will be seriously injured in a collision will grow progressively greater according to whether he hits a motorbike, a 'mini' car, a Rolls Royce, or a large lorry. A lorry is not inherently any more likely to contribute to the occurrence of road accidents than is any other kind of vehicle; it is more likely, however (again if such factors as speed are held constant), that a car/lorry collision will have more serious consequences than a car/car impact accident. This does not mean that the larger the vehicle involved in any accident the more serious will be the outcome; it is the size and weight differential which is of crucial importance. Lorry/lorry accidents, for example, may lead to less serious injuries than collisions between small and large cars.

Statistical evidence shows, as might be expected, that there is a relatively high probability that car/lorry collisions will result in serious injury or death. One estimate suggests that there is one fatality for every six car/lorry collisions.[5] There is, however, some uncertainty about the exact degree of lorry involvement in death or serious injury accidents affecting car drivers or passengers. A study based on a sample survey of coroners' reports on vehicle collisions in the period October 1967 to March 1968 gives a particularly black picture of goods vehicle involvement.[6] The conclusions of this report include the statement that 'a high proportion (at least 50 per cent) of the car and light van occupants killed in collision are the result of hitting a heavy commercial vehicle'. The actual figures are shown in table 43.

**Table 43 Car occupants killed and vehicles involved: car and other vehicle collisions: Ruffell Smith Survey, 1967–8**

| Type of other vehicle involved in collision | Number killed |
|---|---|
| Car or light van | 117 |
| Heavy commercial vehicle | 147 |
| Public service vehicle | 11 |
| Unrecorded | 21 |
| Total | 296 |

There are two reasons why these figures may tend to suggest that goods vehicle involvement in fatal accidents to car occupants is rather worse than may actually be the case. Firstly the omission of all other causes of death to car occupants, except collisions with vehicles having four or more wheels, can give a misleading impression of the relative importance of lorry involvement. The largest number of fatal accidents to car passengers and drivers in 1968, according to figures in the sample discussed below, occurred when no other vehicle at all was involved in the accident. The second doubt about these sample-survey-based figures is that they do not appear to agree very well with figures recorded in another TRRL study. An analysis of fatal casualties by vehicles involved based on police records in 1968 (excluding accidents involving more than two vehicles) contained the figures for car drivers and passengers shown in table 44. Medium and heavy goods vehicles were thus involved in 23·8 per cent ofthe fatal accidents to car occupants. If all goods vehicles are included this figure becomes 28·0 per cent. Of car occupants' fatal accidents, 33·5 per cent took place when no other vehicle was involved, 32·7 per cent took place in car/car accidents, and 3·6 per cent in car/public service vehicle accidents. It is not easy to compare these figures exactly with those in the Ruffell Smith study because the latter included deaths to the occupants of 'light vans' with the car occupant fatalities. But the coroners' reports used in this study contained terms such as 'Jeep', 'Land Rover' or van and 'lorry' and

**Table 44 Vehicles etc. involved in fatal accidents to car drivers and passengers, 1968**

| Other vehicle, etc. involved | Car drivers killed | Car passengers killed | Total car occupants killed |
|---|---|---|---|
| Pedal cycle | 1 | 4 | 5 |
| Motorcycle | 7 | 2 | 9 |
| Pedestrian | 2 | 1 | 3 |
| No other road user | 310 | 322 | 632 |
| Car | 294 | 324 | 618 |
| Public service vehicle | 46 | 22* | 68 |
| Light goods | 40 | 40* | 80 |
| Medium goods | 29 | 17* | 46 |
| Heavy goods | 233 | 170* | 403 |
| Other vehicles | 10 | 14* | 24 |
| Totals | 972 | 916 | 1,888 |

* In these cases the figures do not distinguish between car passengers and passengers in the other vehicle concerned. In each case it has been assumed that all the passengers were car passengers.
Source: *Contributions to the accident situation by all types of vehicle* (TRRL Leaflet LF 320; reproduced in *Towards safer road vehicles*, TRRL Report LR 481).

'truck', which are imprecise.[6] An approximate comparison of the two sets of data is contained in table 45.

The 'with light goods deaths' included fatal accidents to light goods vehicle drivers and passengers in collisions with cars, other light goods vehicles, medium and heavy goods vehicles and public service vehicles. (In light goods/heavy goods collisions it was assumed that all the deaths were those of light goods vehicle drivers or passengers. With light goods/public service vehicles collisions 6 out of 11 fatalities were attributed to light goods vehicles.) Whichever basis is used, the 'police record' figures show a substantially lower medium/heavy goods vehicle involvement than do the 'coroners' return' figures. The former would seem to be more reliable as they are kept in more detail, and the figures were based on nearly complete returns. It is clearly necessary, however, to have figures available from other years to determine more accurately the true rate of heavy goods vehicle involvement in car occupant fatal accidents.

The Ruffell Smith 'coroners' reports' study contains some interesting information about the nature of some of the collisions recorded. 63 of the car/commercial vehicle deaths occurred in 'head-on' collisions and 37 in head-to-tail collisions (in 10 cases the collision was sideways and in

**Table 45 Comparison of coroners' reports and 1968 police records analysis of car occupant fatalities in four-wheel or more vehicle collision accidents**

| Other vehicles involved | Percentage of total fatal casualties | | | |
|---|---|---|---|---|
| | | Police records | | |
| | Coroners' records % | With light goods deaths % | | Without light goods deaths % |
| Car | 42·5 | 49·0 | 55·8 | 50·9 |
| Light goods | | 6·8 | | 6·6 |
| Medium and heavy goods | 53·4 | 38·5 | | 36·9 |
| Public service vehicles | 4·0 | 5·6 | | 5·6 |
| Totals | 99·9 | 99·9 | | 100·0 |

Source: *Contributions to the accident situation by all types of vehicle* (TRRL Leaflet LF 320; reproduced in *Towards safer road vehicles*, TRRL Report LR 481); H. P. Ruffell Smith, *A study of fatal injuries in vehicle collisions based on Coroners' reports* (RRL Report LR 316, 1960).

37 cases the direction was not known). By comparison 64 of the car/car fatalities occurred in head-on collisions and none in head-to-tail collisions. The relatively large number of fatal accidents in which cars hit the rear of goods vehicles is an important factor in considering the possible remedies for car/lorry collisions. Other studies have found similar results and one study also found that four out of five goods vehicles involved in head-to-tail collisions involving serious injury were stationary.[7] It is also of interest that in the Ruffell Smith study, 'no collisions in which heavy commercial vehicles ran into the rear of cars were recorded.'

## 3 Rail accident involvement

Railways are in some ways intrinsically safer than roads. The main railway advantage in safety is that only the railway undertakings are allowed to operate trains. When the railways were first developed in the early part of the nineteenth century it was at first thought that the rail companies would provide the track and allow anyone who wished (and who would pay the tolls) to run trains on it. This proved impracticable and the exclusion of the general public from running their own trains provides a great contrast with road operation. Greater rail safety has little to do with the actual technique of using iron tracks instead of roadways. If all private traffic was excluded from the roads and only scheduled road trains were operated, then the safety record of road transport would probably be at least as good as that of rail. It can indeed be argued that, technically, road operation is potentially safer than rail operation since road vehicles can avoid each other in a way which rail trains cannot. Rail safety requires elaborate and expensive signalling equipment and signal operators. Present rail safety levels could be affected adversely if the railways were used more intensively or if rail speeds were increased significantly.

At present speeds and levels of operation, however, the accident rates for rail operations are low. The Independent Commission on Transport[1] estimated road and rail fatality and serious injury rates per thousand million passenger miles travelled and produced the following figures (for 1971):

| | |
|---|---|
| Road deaths | 31·0 |
| Rail deaths | 8·6 |
| Road serious injuries | 367 |
| Rail serious injuries | 12·4 |

These figures exaggerate the rail advantage since rail, bus and lorry crew miles travelled were omitted from the denominators in the calculations, but all rail and road deaths were counted in to the respective numerators. This does not put road and rail on a comparable basis (as was suggested in the Independent Commission Report), since the road miles travel figure omitted would be much larger. Total loaded rail freight trains miles run in 1973, for example, amounted to approximately six million,[9] whereas total lorry mileage in 1967–8 was about 15 thousand million.[10] Even allowing for a larger train crew, and for rail empty running, it is evident that there are many more miles travelled by lorry drivers than there are by freight train crews. Adding the respective sets of figures into the denominators in the Independent Commission's calculations would reduce the rail safety advantage, but rail would still appear as the safer of the two modes. Another calculation,[11] for an unspecified European country (probably France), produced the comparisons:

People killed per $10^9$ passenger kilometres

| | |
|---|---|
| Rail | 4·5 |
| Road | 9·8 |

People injured per $10^9$ passenger kilometres

| | |
|---|---|
| Rail | 2·6 |
| Road | 204 |

Lower rail accident rates per passenger mile (or per ton mile) do not necessarily imply that accidents could be avoided by moving goods or passengers from road to rail. As is shown in Chapter Nine, it is seldom possible to replace an all-road journey by an all-rail journey (or haul). If there was a direct rail link from a factory at A to another plant at B, then the probability of accident-involvement would be lower if consignments from A to B were sent by rail. But if it was necessary to send the consignment by road to an inner-city goods station near A and to deliver it by road from another goods station, situated in a built-up area, to B, then the overall accident involvement probability of the road-rail-road trip might be as high, or higher, than that of an all-road delivery. Similarly low accident-involvement probabilities on a passenger train journey could be offset by high accident probabilities in congested streets leading to the rail passenger stations. For passenger travel people can to some extent choose between lower and higher accident-involvement probability modes of travel. An individual can, for example, reduce the probability of his being involved in an accident by making a journey as a bus passenger and then as a rail passenger rather than as a car driver. Whether the state should interfere with people's risk-taking choices is a matter on which value judgments may differ. It is undoubtedly desirable that the information on relative accident-involvement probabilities should be available so that decisions can be made on a rational basis.

## 4 Methods of reducing road accident rates

A TRRL Survey of the causes of road accidents listed three main groups of 'Contributory Factors' and attempted to estimate the percentage of accidents in the survey to which each of these applied.[12] These factors, with the appropriate percentages, were:

| | |
|---|---|
| Human factors such as driver's skill, health, driving too fast, not taking enough care, overtaking improperly | 74% |
| The road environment, design and visibility of junctions, effectiveness of hazard warning signs, condition of road surface | 28% |
| Vehicle design and condition | 25% |

As might be expected, the 'human factor' was shown to be by far the most important cause of accidents. Driving tests already exist for all classes of driver, with especially rigorous tests for drivers of public service vehicles and heavy goods vehicles. High insurance rates, which may be regarded as a manifestation of the pricing mechanism, are a form of discouragement to young potential drivers whose likelihood of accident involvement is high. The 'breathalyser', with the threat of disqualification, has acted, at least temporarily, to discourage drinking and driving. The total driving hours of public service vehicle and goods vehicle drivers are controlled by legislation. To decide whether further state action intended to reduce 'human factor' accidents would be justified is not easy. Periodical second and subsequent driving tests might improve driving standards, but they would be administratively costly, and unpopular. There is no evidence that the reduction of maximum driving hours, for public service and goods vehicle drivers, from ten to eight per day, as required by the EEC, will reduce accident rates. It seems probable that many accidents are still caused by drivers whose judgement has been impaired by alcohol. In 1972, 26 per cent of all drivers killed in road accidents had an alcohol content in their blood of more than 80 mg per 100 ml (the legal maximum). But it is difficult to know what further action could be taken to prevent drunken driving. Society is unlikely to accept a total ban on the sale of alcohol and there is no way in which those who are about to drive can be identified and refused alcohol. There are enforcement problems with all legal controls and there are, of course, many motorists whose blood alcohol content is above 80 mg per ml but who are not detected. Similarly, it is difficult to enforce the 10-hour maximum driving control for lorry and coach drivers. It seems likely that the better enforcement of existing controls would do more to reduce accident rates than would the adoption of more stringent control levels.

There is an EEC Regulation (543/69) which limits daily driving hours for drivers of vehicles of over 3·5 tonnes gross weight to eight hours. This Regulation also limits weekly driving hours to 48 and states that drivers of vehicles of more than 20 tonnes gross weight should not cover a distance of more than 450 km in one day. The Regulation should have been applied to British and Irish domestic traffic on 1 January 1976, but at the insistence of the British and Irish governments this was postponed. Existing maximum daily driving hours are ten in the U.K. and 11 in Eire. The EEC Regulation seems to have been inspired by a desire to limit road competition with state railways, and to prevent the 'exploitation' of lorry drivers, as much as by any belief that accident rates would be reduced. It is obviously possible that drivers may become more accident-prone if their driving day (with breaks) exceeds eight hours, but there is no evidence to show that this is so. Accident risks could be increased if drivers tried to compensate for the shorter working day by driving at higher speeds.

Speed, although not mentioned in the TRRL analysis, can be considered in relation to drivers' behaviour. There is very little doubt that accident rates could be reduced by lowering speed limits. Although it is difficult to disentangle the various factors which cause accident rates to change, the evidence in Britain suggests strongly that the introduction, or lowering, of speed limits has led to falls in accident rates. This was true, for example, of the introduction of the 70 mph limit for motorway traffic. The 1973–4 speed limit restrictions coincided with a fall in traffic (both were a result of the fuel crisis), but accident rates fell more sharply than the decline in traffic. In the quarter April–June 1974 personal injury accidents were 10 per cent lower than in the same quarter in 1973, while motor traffic declined by only 4 per cent. For the whole of 1974, compared with 1973, traffic declined by 3 per cent and injury accidents by 7 per cent.[1] The community could, it seems, reduce accident rates if it was prepared to accept longer journey times. The trade-off between the two 'intangible' benefits of time savings and accident rate reductions creates particularly difficult problems of measurement.

The road factor is clearly important in influencing accident rates. The lower accident rates on motorways (despite higher average speeds) illustrates the advantages of eliminating intersections and crossings, and of removing most of the danger from overtaking. Motorway accident rates are lower than those on all-purpose roads for vehicle users. The comparison would be even more favourable to motorways if accidents to pedestrians were included since these are virtually eliminated on motorways. But it is not necessary to build motorways to reduce pedestrian accident rates. Many all-purpose by-passes and other new roads effectively remove much traffic from pedestrians and are thus likely to cut accident rates. As is shown in Chapter Ten, the pedestrianization of shopping areas is also likely to have a favourable influence on accident rates.

Vehicle design was the least important of the three main factors identified by the TRRL analysis. There are some new technologies which can be applied to car design, such as the use of bumpers which throw people colliding with a car free of the wheels. But possibly the greatest immediate scope for improvement in vehicle design is with lorries. It is hardly necessary to stress that high standards of lighting, braking and general vehicle maintenance are essential for heavy goods vehicles. But work on the fairly common head-to-tail collision has shown that changes in the design of the back of some lorries can have an important effect in reducing the severity of such accidents. A particular danger when cars hit the tail of lorries is that the bonnet and front of the car may run under a projecting load, or part of the lorry, the projection then hitting the occupants of the car. According to estimates made by Ruffell Smith it would be worth while to build up the rear of vehicles to prevent under-run. The other obvious measure to prevent head-to-tail accidents is to make the rear of vehicles more conspicuous with extra lights. It is also, of course, essential for lights to be kept clean from obscuring mud.

The main area of uncertainty and controversy in relation to goods vehicle design and accident involvement is the effect of vehicle size and weight. Since a collision with a heavier vehicle is more dangerous than it would be with a lighter one it is not surprising that the possibility of making lorries even more dangerous has been used as an argument against increasing maximum permitted gross vehicle weights. But the position is by no means clear. During the whole of the period 1960–70 lorries of 'over 1·5 tons' ULW (unladen weight) had a lower accident involvement rate than lorries and vans under that weight. This may partly be because the lighter vehicles operate more frequently on urban roads where accidents are more common. The danger of the heavier vehicles, as has already been argued, is not greater accident involvement, but involvement in more serious accidents. The statistical evidence available, however, is quite inadequate as a basis for determining whether the severity of accidents continues to increase with gross vehicle weight over all weight ranges. Most data only distinguishes between over and under 1·5 tons ULW, or at best over and under 3 tons ULW. Possible increases in gross vehicle weight relate to vehicles of more than 10 tons ULW. There may well be a point beyond which further weight increases will not affect the severity of an impact accident. Whereas it may be more dangerous to have a head-on collision with a double decker bus rather than a saloon car, there may not be very much difference in outcome between hitting a 32-ton or a 40-ton lorry. (It is assumed that increases in vehicle weight will not be made at the expense of braking efficiency or ease of control of the vehicle.) If an increase in gross vehicle weight does bring some potential increase in the severity of injury accidents, then the vital question is how far this would be offset by having fewer lorries on the roads. There are some types of accidents which, it seems likely, would be reduced in frequency if fewer larger vehicles replaced more slightly lighter lorries. Overtaking accidents would be reduced if there were fewer slow-moving lorries to overtake (and fewer lorries to hit when carrying out a risky overtaking operation). The gains from having fewer lorries to hit in head-to-tail collisions would almost certainly offset the disadvantages of the lorries being heavier within the likely range of increase from 32 to 40 tons). On the other hand a heavier lorry which goes out of control may do significantly more damage than one which is a few tons lighter. There is also a problem with lorries which carry dangerous loads where the potential damage may increase proportionately with the weight of the load. It may be that, where it is impossible to avoid sending dangerous loads by road, the size and weight limits should be lower than for other loads. (This restriction already applies to the carriage of petroleum.) But this, of course, leads to the familiar dilemma of whether more small dangerous loads would be better or worse than fewer, larger ones.

In the present state of knowledge all that can be said with conviction about the relationship between lorry weights and road accidents is the negative statement that it is certainly not true that an increase in lorry weight would necessarily make the accident situation worse. The best guess is that heavier and fewer lorries would mean fewer lorry-involved accidents, but that a small proportion of the accidents which still took place might be more severe. If permitted vehicle weights were reduced in some urban areas (other factors remaining unchanged) then the laws of probability suggest that there would be more accidents as a result of the increased population of lorries. Some accidents might, however, be less severe, so that there could be a trade-off between having fewer more serious accidents, and more less severe ones.

## 5 The cost of accidents

The priorities and assumptions of the community in relation to the problem of transport accidents are expressed by giving a money value to the 'cost' of accidents. Occasionally it may be possible to put forward a policy choice as a relatively simple selection between two alternatives. Thus it might be shown that an x per cent cut in speed limits on all-purpose, two-way roads would cause a y per cent reduction in accident rates. The community, or its elected representatives, could then make the decision about which benefit was valued most, the shorter journey times or the safer roads. But numerous decisions on, for example, possible investment to install a stretch of street lighting, or to improve an accident 'black-spot', cannot all be referred directly to the arbitration of public opinion. Decision-takers must have a rule to work by, and the only practicable one available is to compare the money value of benefits and costs. Ideally the costs and benefits of alternative accident prevention measurements should be compared with each other, and also with those of other methods of reducing human physical suffering, such as investment in hospitals or in attempts to suppress the smoking of tobacco.

It is not proposed to give a detailed discussion on the difficult problem of evaluating the benefits of accident prevention (or the costs of accidents), which has given rise to an extensive literature.[13] Estimates of the costs of accidents in Britain include allowances for both 'measurable' and 'subjective' costs. The 'measurable' or narrow economic costs include those of damage to vehicles and other property; the services of the police; medical treatment; legal action; insurance administration; and lost output. For fatal accidents to people still at work the main 'measurable' cost is the discounted value of the output which the person killed would have produced in the rest of his working life if the accident had not happened. In the original paper by Dr R. Dawson which produced 'official' estimates of the costs of accidents the discounted value of what a person would have consumed if he had lived was deducted from the costs.[14] This produced the inconvenient result that, before any subjective sum was added, the value of saving the average female life was negative. The low money value of housework gave a lower discounted value for output than for consumption, so, from what Dawson called 'a strictly material point of view' society would gain when the average woman was killed in a road accident. In this paper Dawson added a fairly arbitrary sum of £5,000 as the 'subjective' value of life which made all lives worth saving. This subjective sum covers the value of the emotional loss to the victim himself, and to his family and friends, of a premature death in a road accident. (There are also subjective costs of injury accidents.) The arbitrary nature of this important 'subjective' value of saving accidental death or injury makes it very difficult to carry out any convincing cost benefit analysis on accident prevention measures. Some people would make the value judgment that much higher values should be used. The little evidence which exists on what people or their family would pay to avoid death or injury is not very helpful. The relevant actions taken by individuals which can be observed relate usually to small changes in the probability of suffering in an accident. Methods of evaluating life based on what people will pay to avoid slightly higher accident risks have not yet produced satisfactory figures of the average subjective value of life which can be used in decision-making in the real world.[15]

Since Dawson's original estimates a change of philosophy has been accepted officially which considerably increases the 'cost' of fatal accidents. The 'net value of output' approach, in which consumption is deducted from output, has been replaced by 'gross value of output' estimates in which no deduction is made. It is now argued that a person whose life may be saved by accident prevention methods should be considered to be a part of society, and if benefits are measured after the fatal accident has been prevented, then he is alive to enjoy his consumption and it should not, therefore, be deducted from the value of output.[16] This change had the effect of increasing Dawson's estimate of the average cost of a fatal accident from £9,270 (1965 prices) to £16,980 (1968 prices).[17] The 'gross value of output' approach is based on the assumption that the 'cost' of a life is what a man would himself pay to avoid a fatal accident, and it appears to be illogical to add a further 'subjective' value when consumption is not deducted from output. Dawson himself suggested that his revised figures may involve some double counting.

The average costs of accidents in 1974 for all roads (based on Dawson's analysis) were assumed to be:

| | £ |
|---|---|
| Fatal | 30,000 |
| Serious injury | 2,200 |
| Slight injury | 400 |

The total cost of road accidents in Great Britain in 1973 (on the basis of the 1973 cost figures) was estimated to be £615 millions.[2] These accident cost figures influence decision-making in new road investment since they are used as the basis for evaluating the benefits (or costs) of any expected change in accident rates. Accident rate changes do not have any great influence on the figures in any cost benefit analysis of a major new road project as the values are usually very small in relation to those created by reducing journey times. For small-scale improvements of accident 'black-spots' or 'black sites' a different, and very unsatisfactory, method is used by local authorities (under guidance from the Department of the Environment). Action to improve the conditions on high-accident risk roads are only allowed when accident rates reach a specified level. Thus one criterion currently in use for rural local authority roads is that improvement to a particular stretch made in order to reduce accidents will only be allowed when accident rates exceed 0·75 per million vehicle kilometres per year. An alternative standard is that at least seven accidents must have been recorded during the last three years.[18] Financial control in Nottinghamshire implies that 'only individual [road safety] schemes having a predicted first year economic rate of return of 100% may be implemented'.[19] Given that there are a very large number of stretches of road in Britain where safety conditions could be improved, and on the assumption that all these roads cannot be improved simultaneously, then there is an obvious logic in giving priority to the improvement of those stretches of road with the worst accident records. But it is quite unsatisfactory to use a rigid criterion which involves waiting for accidents to occur before action is taken. It is not impossible to predict which roads will become future accident 'black sites', as a result of changes in the road system and in traffic flows. Local authorities should be encouraged to invest in preventing the development of 'black sites', as well as in removing those already existing. It is quite possible that an alteration in traffic flows will mean that a road which has had a low accident rate in the past can be expected to produce a higher future accident rate than that of an existing 'black site'. It is not sensible to allocate investment intended to reduce road accidents solely on the basis of past records.

It is not at all clear that current road investment is adequate even for improving known existing 'black-spots'. Thus Warwickshire, for example, records 39 'seven plus' accident black-spots existing on rural roads in 1973–4, but how many of these will be improved in their current investment programme seems far from certain.[18] Nottinghamshire identified 84 sites having four or more injury accidents in 1974, but during 1974–5 remedial schemes were introduced at only 17 'black sites'.[19]

Current policy suggests that local authorities should aim to reduce accident rates by 12·5 per cent. This is based on experience in Hertfordshire, the Greater London Council, and the former Road Safety Units of the Department of the Environment, which suggest that a 12·5 per cent reduction can be achieved reasonably cheaply but that beyond this point 'the remedial measures become more complex and hence more costly and the economic return in accidents saved is progressively reduced'.[20] This same magic figure of 12·5 per cent appears in the Nottinghamshire Transport Policies and Programme for 1976–7. The Submission states that 'It was considered that a reduction below that figure (12·5 per cent), although technically possible, would involve diminishing returns and a high level of financial commitment'.[19] It was not made clear whether this figure was discovered independently by Nottinghamshire or based on received wisdom from the Department of the Environment. An effective challenge of the '12·5 per cent' figure would need to be based on independent research. It is difficult however not to be sceptical about the meaningfulness of such a precise figure. It is also hard to believe that the accident situation is identical in all parts of Britain. The 'cut-off' point must in any case be influenced by the assumptions made about the value of preventing accidents. However, it is probable that all that the 'target' actually implies is that the more expensive accident prevention methods will not, at present, be undertaken.

## Summary

*1. Considerable care is required in interpreting accident statistics. Accident involvement does not necessarily imply any degree of responsibility. Some vehicles have high accident involvement rates because they run mainly on high-accident-risk roads. It is also necessary to distinguish between situations in which the vehicle driver is the person most likely to be injured in an accident and those in which it is more likely to be another road user.*

*2. In relation to distances travelled, the vehicles least likely to be involved in accidents of all severities have been lorries of over 1·5 tons unladen weight, and those most likely to be involved have been motorcycles. 1973 figures show motorcycles being involved in over 12 times as many accidents per vehicle mile as the heavier lorries and over ten times as many as cars and taxis.*

*3. For fatal accidents cars and taxis had the lowest involvement rate while that for motorcycles was again the highest. Public service vehicles have the highest rate of involvement in pedestrian fatalities.*

*4. The severity of vehicle to vehicle accidents is likely to be greater when there is a large weight difference between the vehicles involved. Heavy lorries are involved in a considerable number of accidents in which car drivers or passengers are killed.*

*5. Rail transport in Britain generates less accidents per mile than any form of road transport. But this better safety record may not apply to comparisons between all-road and road-rail-road journeys or hauls.*

*6. Possible measures to reduce road accident rates include the lowering of speed limits; further restrictions on the sale of alcohol; the introduction of periodic driving tests; the allocation of more funds to the improvement of accident 'black site' roads; and changes in vehicle design.*

*7. Methods of measuring the costs of accidents are very imperfect. The 'subjective' cost of accidents is a largely arbitrary figure. It is quite uncertain how far the amount of money invested in accident reduction reflects the actual preferences of the community.*

*8. It is unsatisfactory that investment to remove accident 'black sites' can only be made on the basis of past records, and that a potential 'black site' does not qualify for investment.*

*9. Local authority investment in accident reduction is influenced by such Department of the Environment guidelines as that implied by the 12·5 per cent reduction target. These guidelines should be open to challenge. There is a danger that investment priorities will be decided by processes which cannot be explained to councillors or electors. It may well be that current cuts in expenditure on accident reduction would not be acceptable to public opinion if it was informed about the available options.*

Chapter seven

# Vibration, delays, and visual intrusion

## 1 Introduction

In this Chapter four relatively minor social costs of transport, mainly associated with heavy goods vehicles, are considered briefly. These are vibration; delays caused to other road users by slow-moving vehicles; visual intrusion; and direct damage to buildings. An important social cost of transport is the severance of communities which results from the building of trunk roads, motorways and railways. This problem is discussed in Chapters Eight and Nine.

## 2 Vibration

The main possible adverse affect of vibration is that it may do damage to buildings at the side of heavily trafficked roads. The Royal Institute of British Architects in a memorandum originally produced in relation to a discussion on raising gross vehicle weights and axle loads in March 1970, asked, very reasonably, for a detailed cost benefit analysis of the proposals to be undertaken. The memorandum also argued that the vibration effect on buildings would be increased even if maximum weights were raised but axle loadings remained constant because 'Five to seven feet down – probably at foundation level – a pair of axles each carrying 10 tons will exert pressure of 20 tons. . . . Twenty feet down the whole load on the vehicle would be operative as a vibrating source and it would be this which would affect buildings, not individual axle loadings.'[1] In a letter sent to all professional bodies representing architects practising in EEC countries in January 1973 the President of the RIBA (as reported in the *Guardian*) appeared to take a rather stronger line on vehicle weights, arguing that 'the architectural profession is opposed to the relaxation of these controls because, from its practical experience of designing for the built-up environment, it is felt that such a move would be wholly damaging'.

There are two types of damage to buildings which can be caused by vibration, and it is important to distinguish between these. 'Architectural' damage relates to cracking of plaster and other brittle material, whereas 'structural' damage implies that the building itself is in some kind of danger.[2] Architectural damage may be very annoying to house owners but does not endanger the continual functioning of the building as a house, office or workplace. There are a considerable number of factors which can cause building damage including the effects of weather conditions, settlement of foundations, shrinkage of materials, and deterioration of materials. It is very difficult therefore to isolate the effect of traffic vibration on damage to buildings from that of other causes. The evidence summarized in the survey by Whiffin and Leonard suggests that 'In the case of traffic, the most severe vibrations generated are unlikely to cause direct failure, and it is considered that their main action is, therefore, that of "triggering" (but this might equally arise from door-slamming)'.[2] People will generally feel the effects of vibration directly before the level at which minimal architectural damage is caused. In reviewing the many complaints received about vibration damage the Civic Trust report on heavy lorries considered that some of these might result from attributing damage to traffic which was caused by other factors.[3] The Whiffin-Leonard survey also argued that 'Attention can be drawn to vibration by the rattling of doors, windows, lids of ornaments, mirrors, etc. These objects are often balanced in a state of neutral equilibrium and readily respond to very low levels of vibration. The association of these audible and visible signs, with the possibility of damage to the building, results in exaggerated complaints about vibration even though, in fact, there may be no risk of damage.'[2] The rattling of doors and windows may, however, be held to constitute a nuisance in itself. Although normal buildings may be relatively little affected by traffic vibration the effect on old buildings may be more serious, although this appears to be an area of uncertainty. There may be a need to take special measures to protect historic buildings which could be exposed to serious traffic vibration.

Vibration is affected by dynamic axle loads or wheel loads so that increasing axle loadings of heavy vehicles can have an adverse affect on vibration damage. There is, however, no evidence available which would make it possible to quantify the relationship between axle loads and vibration-induced building damage. The vibration effect of axle loadings can be affected by the nature of the vehicle suspension, and small increases in axle loads could probably be offset by improved suspension systems, although this also is an area of uncertainty. Much more important is the influence of the nature of the road surface. Vibration from road traffic is a result of 'fluctuations of wheel contact load which are caused by their passage over road surface irregularities or discontinuities in the road profile'.[2] As Whiffin and Leonard argue, the best way of minimizing traffic-induced vibration would appear to be the maintenance of smooth road surfaces.

## 3 Delays to other road users

One of the main sources of frustration on many roads in Britain is the delay caused by slow-moving vehicles. To drive along the 42 miles of the A30 trunk 'primary' route between Exeter and Launceston, for example, is almost always a frustrating and, for some personalities, an enraging experience. Cars may be delayed by heavy lorries or by caravans over almost the whole stretch of the road, and speeds can drop below 10 mph on some of the hilly sections of the road. This is, of course, only one example of many similar lengths of road in Britain. The cost to the community must be measured not only in lost time but also in frustration, and to some extent in the incidence of overtaking accidents. One remedy which has been proposed is the imposition of minimum power-to-weight ratios on goods vehicles so that they would be able to travel faster, particularly on gradients and when moving from rest. A cost benefit analysis of the power-to-weight proposals was carried out by the Transport Research Unit at Leicester in collaboration with the TRRL. The main results are summarized in table 46.

In October 1971 the Minister for Transport Industries announced that all goods vehicles used after 1 April 1973 must have engines with an 'as-installed' power equivalent to at least 6 BHP per ton. As the figures in table 46 indicate, no very dramatic improvements in vehicle speeds can be expected from this move or even from a further possible increase to a minimum level of 8 BHP per ton. The majority of new heavy vehicles already have engines giving 6 BHP per ton although many older vehicles may have lower ratios, especially if they have been poorly maintained. It is not possible to achieve more than relatively minor increases in the speeds of heavy goods vehicles on gradients (and in accelerating from rest) by any feasible increases in engine power. There are limits to the power which can be developed from a diesel engine in its automotive form, and alternatives such as the gas turbine are not yet in general production. There are also several environmental problems involved in increasing engine power and vehicle speed. The figures in table 46 were based on the assumption that power-to-weight ratios of 6 or 8 BHP per ton could be achieved without any significant cost in increasing the other social costs of vehicle operation. But engine power is, of course, related to noise and the emission of pollutants, and even an increase to 8 BHP might make it more difficult to achieve lower levels of emission or lower noise levels. Further increases in engine power beyond 8 BHP would not only cost more (mainly in increased fuel consumption and capital cost) than the value of likely benefits but would probably worsen the noise and emissions problems. The present fuel tax and licence excise duty both tend to make manufacturers build low-powered engines which have a low fuel consumption and which are relatively light.

**Table 46 Estimates (less tax), annual costs and benefits of minimum power-to-weight legislation**

Change from 5 to 6 BHP per ton

| Gross vehicle weight (tons) | Social benefits £ | Private benefits £ | Total benefits £ | Costs £ |
|---|---|---|---|---|
| 24/25 | Nil | Nil | Nil | Nil |
| 32 | 86–8 | 48–51 | 134–9 | 40–62 |
| 42 | Nil | Nil | Nil | Nil |

Change from 6 to 8 BHP per ton

| Gross vehicle weight (tons) | Social benefits £ | Private benefits £ | Total benefits £ | Costs £ |
|---|---|---|---|---|
| 24/25 | 62–3 | 45–74 | 107–37 | 80–195 |
| 32 | 62–3 | 45–74 | 107–37 | 282–315 |
| 42 | 62–3 | 45–74 | 107–37 | 213–75 |

Source: C. Sharp and A. Jennings, 'More powerful engines for lorries?', *Journal of Transport Economics* (May 1972).

Increased engine power can thus only provide a very partial remedy for the delays caused by slow-moving goods vehicles. Power-to-weight ratios could be increased by lowering the maximum weights of vehicles, but the costs falling on the community as a result of this measure could be astronomically high, and motorists would in any case tend to lose as much from the consequent multiplication in the number of lorries as they would gain from increases in speed. Cars drawing caravans may be a major source of delay, particularly, of course, on some holiday routes in the summer. The 'power-to-weight' problem could be overcome by introducing rigid rules on the size of car allowed to pull caravans of defined weight. But this would not do a great deal to remove delays since the main effective restrictions on the speed of car-pulled caravans are based on safety considerations. As is the case with lorries, cars with caravan trailers can normally operate at the maximun legal speed (of 40 mph) on reasonably flat roads. A tax representing the costs caused to other road users could be imposed on caravans but this could prove very unpopular politically. Learner-drivers, another common cause of delay, are already banned from motorways, presumably for safety reasons (slow-moving vehicles do not cause delay on uncongested motorways) and the ban could be extended to other busy roads, or to all urban roads at peak traffic flow periods. As is so often the case, however, environmental aims may conflict. Drivers must learn to negotiate busy roads, and they could not be regarded as fully trained if they learnt to drive only on low-traffic-flow routes. The only complete answer to the delay problem is to provide roads on which overtaking can be carried out safely. Where this is not possible the mixture of traffic trying to move at different speeds must cause some delays. One possible future development could be a change in the policy of car manufacturers which would affect the 'demand' to overtake. At present many cars are still designed to travel at speeds which are considerably higher than the legal maxima applying in most countries. A recent car test (of the Aston Martin V8) praises its ability 'to attain 120 m.p.h. in less than half a minute'. It is still thought worth while to claim in an advertisement for the Jaguar XJS that 'The top speed, where permissible, is in the region of 150 m.p.h.'. Safety and fuel conservation considerations may persuade governments to reduce the number of roads on which such high speeds are allowed and this could modify the demand for differential speeds on public roads.

## 4 Visual intrusion

The 'visual intrusion' of vehicles means that their presence on the roads spoils the landscape, or the outlook from houses, and thus constitutes a social cost. The Civic Trust report includes visual intrusion amongst a list of minor problems[3] concerning lorries, but recently more objections to the lorry have been based on this factor. Thus Sir Desmond Heap, President of the Law Society, in a lecture on 'Ambience and the Environment' given at Manchester University, said that 'One particular form of pollution which can easily, and should speedily, be knocked flat on the head is the vulgar and obscene intrusion of juggernaut lorries into places where they are not wanted and cannot, in any seemly fashion, get.'[5] A paper read to a European Conference of Ministers of Transport Round Table included 'the presence of vehicles themselves in situations where they do not fit into the view pleasingly' amongst the social costs of urban road transport reviewed.[6] Visual intrusion is obviously a highly subjective factor and individual valuations of its cost are likely to vary even more widely than are those of noise. Objections to the mere presence of traffic is likely to be greatest from those living amongst the more beautiful and historic urban landscapes of Britain. If it is one's fortune to live and work in the typical urban landscape of many parts of the industrial Midlands then it is not easy to be greatly concerned about the visual intrusion of vehicles. Where lorries or buses are clearly out of scale with the beautiful buildings in a town the only remedy would appear to be to attempt to find some alternative route for them to follow. The issue of designated routes is discussed in Chapter Eight. There is no obvious way in which vehicles can be made more beautiful or aesthetically acceptable. It would be interesting to find out how far people object particularly to the appearance of lorries or whether they also dislike buses (which may be even more out of scale with an urban landscape) and cars. Areas in which traffic is felt to intrude visually are, of course, likely also to be those in which vehicle noise and air pollution may have particularly undesirable effects. There may even be some double counting if 'visual intrusion' costs are added to noise and pollution costs since dislike of the appearance and presence of traffic may be linked with noise levels and smoke emissions.

Other manifestations of transport activities, besides the vehicles themselves, could also be said to lead to visual intrusion. Petrol stations and garages 'intrude' into many otherwise beautiful small towns and villages. But a discussion of the spoliation of the landscape by buildings would perhaps be better placed in a study of the problems of town and country planning, rather than in a survey of the environmental costs of transport. Roads and railways also intrude into the landscape, but this problem is considered in Chapters Eight and Nine.

## 5 Damage to buildings

The Civic Trust report contains several photographs of buildings which have been damaged by direct impact with heavy lorries. This direct damage to buildings by lorries, though there are no national statistics to measure it, is undoubtedly a real problem. Like 'visual intrusion' it is a symptom of lorries using unsuitable roads and there is no obvious remedy except in the search for, or construction of, alternative routes. Buses and cars also sometimes hit buildings but the damage caused is unlikely to be a function of vehicle weight.

**Summary**

*1. There is little evidence that traffic vibration at its present level causes significant damage to normally constructed buildings.*

*2. Traffic vibration can be an annoyance to people living beside busy roads without causing any 'architectural' or superficial damage.*

*3. Bad traffic-induced vibration may be a factor in 'triggering' such architectural damage as cracks in plaster.*

*4. The effect of traffic vibration on very old structures is not certain but it may be considerably more serious than it is for more recent buildings.*

*5. Vibration increases with dynamic axle loads, and more would need to be known about its effects before evaluating a large increase in axle loads. There is no evidence, however, to suggest that an increase in the range 10–10·8 or 10–11 tons would have a significant effect on vibration-induced damage to buildings.*

*6. Vibration may be reduced by improvements to vehicle suspension.*

*7. The nature of the road surface is a vital factor affecting traffic vibration and improving road maintenance would be the most obvious and effective way of reducing vibration from traffic.*

*8. Increasing the engine power of lorries and buses can lead to a small reduction in the delays caused to other road users by slow-moving vehicles. But there are engineering limitations on the power of diesel engines, and large increases in power would involve considerable private costs (fuel consumption and capital costs) and some social costs (increased noise and emission levels).*

*9. Major reductions in the delays caused by heavy goods vehicles can only come from road improvements and new road building. The 'demand' to overtake could be reduced by lowering the top speed of cars.*

*10. It is difficult to find any solution for the problems of 'visual' intrusion and direct damage to buildings except in searching for, or constructing, alternative routes which would remove heavy traffic flows from unsuitable roads.*

hapter eight

# Roads and the environment

## 1 Introduction

The environmental costs of road transport would be reduced virtually to zero if transportation could be carried out at a distance from people. There are very many industrial processes which would be much more offensive than transport operations if they were carried out in the High Street or on the village green. But manufacturing activity is largely confined within factory walls, or inside the fence round an industrial plant. Modern industry is generally located in zones where its impact on people is minimized. Transportation must, inevitably, be carried on where people live. One of its main functions is to bring goods to people, and to move people themselves. Another function is to link industrial areas, and to do this without traversing populated country may be impossible. But if a total separation between transportation and people is impossible, very considerable environmental gains may be achieved by organizing a greater degree of separation than at present exists.

Although this Chapter has 'Roads' in its title, as it deals mainly with road traffic, it might have been given the more general title 'Routes'. If major flows of traffic can be concentrated on certain routes, whether the mode of transport is road, rail or air, and if these routes can be made to go through largely uninhabited areas, then the environmental costs will be reduced to a minimum. Nobody will be greatly concerned about the noise, localized air pollution, vibration or visual intrusion of heavy traffic flows along a road which is built through a desert. Although there are no deserts in Britain there are areas where the environmental costs of traffic flows will be very low, and others where they may be extremely high. The aim of the 'road' solution to the problem of transport environmental costs is to move traffic flows from high cost to low cost areas. When this can be done it has the great advantage of dealing with all the main environmental costs, except accidents to road users, at once. If there was any immediate likelihood of building major new railway lines or canals in Britain, then it would be necessary to discuss the problem of their environmental impact. But as it is only new roads that are likely to be built in the near future, we concentrate here on this transport mode. The relative environmental advantages and disadvantages of existing rail routes are discussed in Chapter Nine. In this Chapter the ways in which environmental factors influence decisions on new road investments are examined, and the case for building new roads in order to improve the environment is analysed.

## 2 Environmental factors and road investment appraisal

The problems of evaluating highway investment form a major study on their own. The theoretical issues and their application to road and other forms of transport investment are clearly described by A. J. Harrison in a recent book.[1] The Department of the Environment, the Scottish Development Department, and the Welsh Office are responsible for investment in trunk roads and motorways in England, Scotland and Wales respectively. At the time of writing the Scottish Development Department and the Welsh Office broadly follow similar investment appraisal procedures to those developed by the Department of the Environment (although there is some evidence to show that the 'development' factor has been given greater weight in Scotland).[2] All other roads are now the responsibility of the new Shire or Metropolitan County Councils.

It is only relatively recently that formal methods of road investment appraisal of any kind have been developed in Britain. Even when the earliest motorways were built no attempt was made to compare benefits expected from alternative projects. The most obvious example of this was the construction of the M50 from Strensham (in Worcestershire) to Ross (in Herefordshire) as one of Britain's first motorways. No formal investment appraisal of this motorway was attempted when it was decided to build it in 1957.[2] The plans were prepared by enthusiastic local authorities, it was considered desirable to provide a road which would eventually form part of a link between Birmingham and South

Wales, and so the motorway was constructed. But, judged on the basis of any conceivable economic criteria, there were very many other places in Britain which should have been given a motorway long before this part of the West Midlands. Even today, when the 'Ross spur' does link up with the main motorway network of Britain, its traffic flow figures are very low by motorway standards. The average 24-hour traffic flow on M50 in 1971 was 8–9,000 vehicles, whereas average flows on stretches of M6, which was built much later than M50, ranged from 26,000 to 37,000 vehicles per 24 hours.[3] Another obvious error in early motorway planning was the reduction of M1 from three lanes to two lanes as it entered the London area beyond the M10 junction. This curious result of the lack of detailed investment appraisal and the absence of accurate traffic forecasting was described as 'the most amazing folly' by a member of the House of Commons Select Committee which investigated the motorway and trunk road programme in 1969.[2]

During the 1960s, and particularly during Barbara Castle's term of office, the Ministry of Transport developed techniques for comparing the expected benefits from alternative road investment projects. The basic measurements were of the costs of road construction and of the benefits from what has, rather obscurely, been called 'Travel and Accident Loss'. This covered changes in accident rates, in the costs of vehicle operation, and in travelling time. Allowance was also made for changes in road maintenance costs. This technique has been used in various more or less sophisticated forms, but the 'costs' and 'benefits' measured are still essentially the same today. The Ministry of Transport and their successors the Department of the Environment have attempted to develop road investment appraisal with the 'COBA' system into what they have sometimes been bold enough to call 'full cost benefit analysis'.[2] The main developments have been the measurements of the effects of a road investment over the whole network of roads affected (and not just those of the new road itself), and an extension of the time period over which costs and benefits are estimated. Both these developments are obviously desirable. If a new road is built, or an existing road improved, some of the benefits will take the form of the reduction of congestion on other roads from which traffic will be diverted. In an urban area, new road construction may actually reduce speeds on some roads. The redistribution of traffic resulting from the completion of the new road may mean that some unimproved roads in the urban network will carry more traffic than before, so that average speeds will be reduced. The practice of estimating the benefits of a new road only for the first year of its life (usually called the 'first year rate of return') is also clearly less satisfactory than estimating benefits over a period of 30 years (as is now recommended by the Department of the Environment). The traffic flow on some new roads may be expected to build up over time, whereas that on others may reach its peak level almost immediately. First-year rate of return criteria would bias investment decisions in favour of the latter type of road.

Before discussing the shortcomings of Department of the Environment road appraisal methods in making allowance for environmental factors, we should consider the very considerable difficulties involved in applying even the existing techniques. All the rate of return calculations are based on traffic flow forecasts and these are not at all easy to make. The problem is relatively simple for a major motorway or trunk road connecting two urban areas, though even here there must be uncertainties about what will happen in 20 or 30 years' time. Variations in government regional economic policy, changes in the availability of fuel for cars and lorries, and the development of alternative non-road forms of transport are just three of the factors which could affect the future size of traffic flows. In urban areas, where traffic flows through a network of roads, forecasting is much more difficult. In order to extimate the future traffic flows on one road in a town it may be necessary to make assumptions about what other road improvements will or will not take place during the period of the forecast over the whole of the local road network. In Leicester, for example, estimates of traffic flows on inner city roads will be greatly affected if some proposed outer ring roads and relief roads are built. But at present it is not possible to say when these roads will be constructed or indeed whether they will ever be built at all.

A second major stumbling block in the way of applying existing road investment appraisal techniques is that of calculating a rate of return on the cost of building the road, which involves putting money values on those benefits which are measured. If Britain had toll roads then it would be possible to find out how much people would be willing to pay to run their vehicles on them, but as no direct charge is made for using the roads this information is not available. The most important benefit to result from new road construction or improvement will usually be that journey times will be shortened. But putting a money value on time involves as many difficulties as does the evaluation of environmental benefits. There is a formidable amount of literature devoted to the subject of the value of time in transport investment.[4] Since time values are so important in most transport investment project appraisals, mistakes in time evaluation can completely 'swamp' the effect of the inclusion of environmental factors in the appraisal.

The only environmental factor directly included in the 'COBA' benefits is improvement in accident rates. The problems of evaluating the costs of accidents have been discussed in Chapter Six. Some environmental factors are included, indirectly, on the 'cost' side of the calculations.

The estimated cost of constructing a road (the relation of these costs to the value of expected benefits determines the 'rate of return') will include the cost of the land which will have to be acquired and that of any buildings which it will be necessary to buy and demolish. So long as land and buildings are bought at fair market prices this should cover their alternative use value and the objection that 'roads use up precious agricultural land' becomes irrelevant. If the community put a higher value on a strip of land in producing wheat than in carrying a motorway, then the market price of the land would be too high for a proposed road built on it to show a satisfactory rate of return. There are two problems in relation to land purchase, however. As already mentioned, road building could sometimes involve the destruction of beautiful or historic buildings whose 'fair market price' is very difficult to determine. Secondly, the market price of a house which is acquired compulsorily may not cover the loss to people who are forced to move away from their old homes, and some extra compensation may be required. Perhaps the main objection to road investment appraisal techniques has been that no direct allowance was made for the adverse effects of new or improved roads on nearby property. Houses which were near to a new road (but not so near that they have to be acquired by the road builders) might, for example, suffer from greatly increased traffic noise levels, or by being cut off from a part of their local community by a busy trunk road. Road investment appraisal clearly falls short of full cost benefit analysis if it ignores such relevant social costs as those of people kept awake by traffic noise or children barred from playing with their friends by speeding cars and lorries. The only way, in the past, in which allowance has been made, indirectly, for the social costs of traffic has been by setting required rates of return for road projects at much higher levels than those which might be expected from alternative non-road forms of investment. A rate of return of 20 per cent has commonly been taken as a minimum acceptable level for a road project.

The government now propose to take some notice of the problem of those whose property suffers as a result of new (but not past) road building by paying limited compensation to them. This course of action was recommended by the Urban Motorways Committee which reported in 1972. The report argued that social as well as private costs should be taken into consideration in new road building and that these costs should include those of traffic noise, severance of communities, the disturbance caused by road building, and the visual intrusion of the road. Following discussion in a White Paper the principle of compensation was introduced by the Land Compensation Act, 1973. Under the provisions of this Act compensation must be paid to those whose property suffers a depreciation in value because of works to construct a new highway or airport. The 'physical' factors mentioned in the Act which may cause a nuisance included noise, vibration, smell, fumes, smoke, artificial lighting and the discharge of solid or liquid substances.[8] Compensation for the first £50 of damage suffered was excluded, and provision was made to offset increases in property values caused by the new works (resulting from improved accessibility for example) against losses caused by environmental deterioration. It is not yet possible to analyse the effects of this Act. Many of the details were left to be worked out in government regulations. Regulations have been made relating to compensation for increased noise levels (the most easily measured factor) and measurements have been taken of the noise levels affecting houses near new by-passes and other road works.

It has been estimated that the costs of the implementation of the Land Compensation Act will amount to about £50 million a year by central government and £12 million a year for local authorities. About half of this sum will relate to remedial acquisition and extra works and half to compensation.[6]

The Land Compensation Act is clearly a step forward if it results in bringing more environmental factors into investment appraisal calculations, though it is uncertain whether administratively determined costs will bear any very close relationship to the actual social costs. In one respect, however, the provisions of the Act could distort investment appraisal. If a proposed new road is to be debited with the social cost which it imposes, it should also be credited with those which it reduces or removes. There is at present no formal method of making allowance for the reduction of social costs which may result from the building of a new road (with the possible exception of some local authority roads discussed below). Suppose that a new by-pass which runs near to 50 houses removes a traffic flow from a stretch of road bordered by 1,000 houses. It is obviously absurd to make allowance for the environmental losses of the 50 houses on the by-pass while ignoring the environmental gains accruing to the occupants of the 1,000 houses on the by-passed road.

The more popular version of the Department of the Environment 'COBA' programme admits that it does not 'include environmental costs and benefits except in as much as a different alignment, or the greater use of cuttings may be adopted for a scheme for such reasons and therefore feature in its assessed costs and benefits'.[9] But since 1973 the Department of the Environment has developed a system of public participation in which the public are supplied with details of alternative routes for proposed new roads and can argue for the route which they prefer (or against the one which they most dislike). Where any local authority objects to the route of a new trunk road or motorway the Minister is also obliged to institute a public inquiry at which objectors often seek to argue not only against a particular route but also against the need for the road to be built at all.

Under the Local Government Act of 1974 many changes were made in the nature of local authority responsibility for road investment. The new shire and metropolitan counties became responsible for road investment, although they could still delegate some powers to district councils. Specific grants for road building and other transport projects were abolished and local authorities must now finance road and other transport investment from borrowing (for which they need consent); from their general rate fund; from the central government Rate Support Grant; or from the new Transport Supplementary Grant. County councils now have to submit 'bids' for the Transport Supplementary Grant (TSG) to the Department of the Environment which specify their proposed expenditure over the next five years. This expenditure covers not only road investment and maintenance but also support for public transport (both road and rail); investment in lighting; expenditure on traffic management; and investment in car parks and lorry parks. Support for public transport can take the form both of grants to cover operating losses and investment in infrastructure. Along with their bid for a Transport Supplementary Grant, shire and metropolitan councils must prepare an annual Transport Policies and Programme (TPP) statement. The TPP sets out the total transport strategy of the local authority. Leicestershire summed up the purposes of their TPP as being to bid for a share of the Transport Supplementary Grant; to explain the county council's plans for local transportation to the Department of the Environment; and to keep county and district council members informed about these plans.[10] A study of a sample of these TPPs gives useful information about the treatment of environmental considerations in making plans for road investment.

Most county councils make some reference to environmental aims in their statements of general policy. The latest Warwickshire TPP states that 'the overall county transportation objective is to allocate resources in order to achieve optimum net benefit to all sections of the community by providing maximum accessibility for persons and freight in ways which are consistent with the conservation and improvement of the environment and minimising the risk of accidents'.[11] Nottinghamshire were even more ambitious, claiming that 'The principal objective of the County Council is progressively to improve the County as a place to live and work in and to visit. Improvement for the majority is not to be at the expense of the under-privileged minority, in fact, there is to be positive discrimination in favour of the under-privileged, and the policy is intended to contribute to an irreversible shift of resources towards those areas, classes, and social groups who have been deprived in the past. The overall County Transport Aim is to strike a balance between the environment, the pedestrian, public transport, the commercial vehicle, and the private car so as to obtain a freedom of movement for people and goods that achieves a buoyant economy and is compatible with a civilised way of life and the minimisation of accidents.'[12] It is no great exaggeration to compare these multiple objectives with those of a First Division football club which stated that its aim for the season was to gain 84 points and win the League and F.A. cups while maximizing revenue from the sale of players, giving due attention to the need for success of all other First Division sides, and minimizing the effort required from its staff. No doubt the county councils concerned realize that these objectives are not much more than a list of desirable ends, and give no guide as to what should be done when objectives are incompatible. Just how a balance can be struck between the claims of the environment, the pedestrian, the commercial vehicle, the private car and the need for a buoyant economy is, of course, part of the basic problem discussed in this book.

The TPPs studied do not contain a great deal of information about the precise ways in which environmental factors are to be taken into consideration in making decisions about road investment. The Warwickshire TPP contained the statement that traffic noise would be used as the main proxy for environmental quality; that atmospheric pollution by vehicles would be considered if feasible monitoring techniques could be developed; and that pedestrian mobility in central areas would be used as a guide to environmental benefits and disbenefits.[11] The Warwickshire document also stated that priority would be given to relieving roads carrying more than 500 heavy goods vehicles per 16-hour day and that 'indicators' of relief would be changes in freight traffic flows, noise levels, and visual intrusion.[11] According to the 1976–7 Kent TPP it was found that there were 300–400 highway schemes which might be candidates for inclusion in the 5-year programme. This TPP contained the explanation that 'As only limited data were available and time would not allow the use of a sophisticated evaluation technique, a simple method based on need was developed and used. This method requires the quantification in monetary units of the safety, economic, environmental and development benefits and the total cost of each scheme. The net benefits are then summed and divided by the cost of implementation to give a value for priority rating'.[13] Assessment of enviromental and development benefits was based on 'subjective judgements'. According to the TPP, 'The local planning division of the County Planning Department suggested values to be used in this submission for environmental benefits'.[13] There were numerous vague references in the TPPs studied to substantial environmental benefits which were used to justify priorities.[6]

A great deal of background information about TPPs and current local authority thinking on road investment was obtained by Mr John Rouse in a short research project carried out as a Visiting Fellow in the Public Sector Research Centre at Leicester University. During this research Mr Rouse discovered that some county councils were interested in developing a 'points' evaluation system for road schemes in which a points currency is used instead of money. Two local authorities in the sample were actually using points systems, although they both said that their schemes needed further development. The description of these schemes here is based very closely on Mr Rouse's work.[6] One of the county councils had a fixed points scale for environmental benefits. Projects are awarded 10 points for a major benefit, 5 for a significant benefit, −1 point for a minor disbenefit and −5 points for a major disbenefit. The project's 'value for money' in creating environmental benefits is then judged subjectively and appropriate multipliers are applied to the benefit points total. Thus a project showing good value for money would have a multiplier of 2. Points are also used for development benefits and to measure changes in maintenance costs. Traffic benefits and changes in accident rates are expressed in money, using 'COBA' figures and methods. The points and money totals are then summed after the application of subjectively determined weights (applied to each major 'benefit' category). The final 'priority' ranking is expressed as a first-year rate of return.

The other county council in the Rouse sample had developed a points scheme for large schemes in urban and inter-urban situations in which the 'COBA' method is almost entirely rejected as a means of determining priorities. The points scoring process is called the 'subjective technical assessment' and the resulting score, together with considerations of financial constraints and spatial equity (which means giving a 'fair' share of investment funds to each part of the county council area) determine relative priorities. The weights given to each aim or benefit of the investment are determined by the county council through the 'local political process'. The objectives, in order of importance and the number of points awarded, are:

Road safety
Environmental consequences

The interests of:
Pedestrians
Public transport users
Commercial vehicle users
Business drivers
Non-peak car users
Commuters

A sophisticated economic priority scheme has been derived to measure the 'road safety' benefits of a project, but for other benefits the evaluation is more subjective. In order to minimize the subjective element an elaborate form has been developed in which the pre-investment problem relating to each 'aim' is ranked as 'severe', 'moderate', or 'small' and the degree to which an improvement will result is listed in such categories as 'substantially' and 'moderately'. On the basis of such an assessment a 16-cell matrix emerges and each cell contains a number of points which are potentially assignable to the project's total score. For example a 'severe' initial problem which is substantially relieved by the project scores 40, while a moderate initial problem which is relieved moderately would receive only 10 points. Apart from benefits, points are awarded for improvements in standards of design (or the effect on maintenance costs), value for money, the political priority given to the major 'project' of which the project under assessment is a part, and finally for 'public pressure' as a measure of the public's attitude to the improvement.

It is not possible to evaluate these schemes without studying their effect on investment decisions over a reasonable time span. In present circumstances county councils may have little effective freedom in allocating their transport expenditure, despite the TPP system. Large commitments to such items as highway maintenance cannot easily be avoided. In tables 47 and 48 the percentages of the Supplementary Grant to be spent on main items of capital expenditure and total transport expenditure are shown.

It is evident that the major commitments to highway investment and maintenance which ranged from 83·7 to 89·4 per cent of the total expenditure of the sample counties (and averaged 86·4 per cent) allowed freedom for only relatively marginal differences in allocation to the other headings. County councils could of course choose different types

**Table 47 Estimated Transport Supplementary Grant expenditure, 1976–7 to 1980–1, six County Councils – capital projects**

| Item | Percentages of estimated Grant — County council: A | B | C | D | E | F |
|---|---|---|---|---|---|---|
| Public transport infrastructure | — | — | — | — | 2·0 | — |
| Bus | 1·9 | 2·8 | 2·6 | 5·9 | — | 1·6 |
| Rail | 3·8 | — | — | — | — | — |
| Highways | 82·3 | 91·0 | 89·3 | 87·9 | 89·1 | 82·2 |
| Lighting | 4·6 | 1·3 | 5·0 | 3·7 | 2·2 | 1·5 |
| Traffic management | 7·4 | 2·0 | 2·5 | 0·3 | 6·7 | 3·3 |
| Car parks | — | 2·0 | — | 1·9 | — | 11·4 |
| Lorry parks | — | 0·9 | 0·6 | 0·4 | — | — |
| Total | 100·0 | 100·0 | 100·0 | 100·1 | 100·0 | 100·0 |

Source: Various county council TPPs.

**Table 48 Estimated Supplementary Grant expenditure, 1976–7 to 1980–1, six county councils – capital projects and current expenditure**

| Item | Percentages of estimated Grant — County Council: A | B | C | D | E | F |
|---|---|---|---|---|---|---|
| Public transport infrastructure | — | — | — | — | 1·0 | — |
| Bus | 0·7 | 1·3 | 1·4 | 2·9 | — | 0·6 |
| Rail | 1·5 | — | — | — | — | — |
| Highways | 32·7 | 41·4 | 47·2 | 43·9 | 43·4 | 31·4 |
| Lighting | 1·8 | 0·6 | 2·6 | 1·8 | 1·1 | 0·6 |
| Traffic management | 2·9 | 0·9 | 1·3 | 0·1 | 3·2 | 1·3 |
| Car parks | — | 0·9 | — | 0·9 | — | 4·3 |
| Lorry parks | — | 0·4 | 0·3 | 0·2 | — | — |
| Maintenance | 50·5 | 46·5 | 38·4 | 45·5 | 42·4 | 54·5 |
| Current lighting | 9·0 | 5·1 | 7·4 | — | 8·3 | 5·4 |
| Miscellaneous current | 0·7 | 2·9 | 1·4 | 4·5 | 0·6 | 1·9 |
| Total | 99·8 | 100·0 | 100·0 | 99·8 | 100·0 | 100·0 |

Source: County council TPPs.

of highway investment. The 'points' systems described both leave considerable scope for subjective judgments, and it is a matter of opinion whether this will give a better or worse result in practice than a full social cost benefit analysis with money values for all costs and benefits. There is an obvious danger of double counting in the points systems, and the local authority using the 'subjective technical assessment' scheme appeared to make allowance for 'political' factors in three places (in determining the initial priorities, in considering the political priority of the project, and in making allowance for 'public pressure').

Even if great importance is attached to environmental factors it does not necessarily mean that they will be decisive in determining the pattern of transport investment. As Harrison has pointed out, competing road schemes, or even competing road and mass transit projects, may differ very little in their environmental impact.[1] Nevertheless the development of independent methods of assessment for investment in roads (and other transport projects) by shire and metropolitan counties could lead to different investment policies and different treatments of environmental factors which varied with counties and which were different from those adopted by the Department of the Environment, the Welsh Office and the Scottish Development Department. The basic priorities of the 'subjective technical assessment' system as applied by the county council using it are essentially different from those used for assessing trunk road projects and embodied in 'COBA'. In 'COBA', as has been shown, time savings are the main determinant of benefits and this implies that the interests of road users have first priority. If the county council using 'subjective technical assessment' keeps to its intention of giving greater importance to road safety and environmental consequences than to the interests of road users then a very different pattern of investment would emerge. Priority for road safety and pedestrians, for example, would imply large-scale investment in pedestrian underpasses and bridges, and in the kind of pedestrianization schemes described in Chapter Ten. What is meant by 'priority' is not always clear, however. Many road investment projects bring large time savings but comparatively small changes in accident rates. If money values are used, as in 'COBA', then projects yielding the greatest time savings in relation to cost will generally be preferred. But a rigid system of priorities would suggest that if there were two alternative projects with identical costs, then the one giving the greatest reduction in accident rates (or improvement in the environment) should be chosen regardless of any other consideration. Nottinghamshire County Council do indeed appear to have made this their policy. Their current TPP contains the statement that 'When considering carrying out roadworks, on whatever scale, priority will be given to those that will achieve the maximum reduction in accidents for a given expenditure'.[12] This implies that all factors other than accident rates should be ignored, and clearly conflicts with trunk road assessment policy.

## 3 Environmental costs and benefits of road construction

There are two main groups of social costs associated with roads, and a failure to distinguish between these sometimes causes confusion in arguments about the desirability of new road projects. Road construction may impose a direct environmental cost on the community if the new road itself spoils a beautiful landscape, or if attractive or historic buildings have to be destroyed so that it can be built. But most of the objections to new road construction are based on dislike for the traffic which is expected to use the road. Opposition to a new road ought, logically, to be modified where it can be clearly shown that the main purpose of the new construction will be to move a traffic flow from where it is causing great damage to the environment to a route where it will be much less harmful.

The environmental costs of traffic have already been discussed in other chapters of this book. There are various direct environmental costs of a new road itself: the social costs associated with the actual process of construction; the impact on the landscape of the finished road; and the loss to the community of any especially significant buildings which have been destroyed to make way for the road. The process of constructing a new road may cause congestion on nearby approach roads, and result in the spreading of dust or mud over a fairly wide area. The road works themselves may be very unsightly, and, like most building operations, look far worse than the finished product. New road building in urban areas may sometimes involve pulling down buildings which have a special value to the community that cannot easily be measured in money terms. The most important environmental cost of a new road, however (as distinct from the cost of the traffic flow which it may carry), is almost certainly the effect which it has on the landscape through which it is built.

Despite the growth of strong objections to new road building it is clear that roads are not necessarily harmful to the landscape. It is difficult to lay down objective standards of beauty but most people would probably agree that some existing roads, with their trees and hedgerows and curving lines can be very attractive, and may improve rather than spoil the scenery. Even modern roads need not do damage to the landscape. W. G. Hoskins, that great expert on the development of the English landscape, while he deplores some motorways as being 'a ghastly infliction on the English landscape', believes that the results of other motorway building have been good, as where the M1 'goes in lovely sweeping curves through Charnwood Forest in Leicestershire'.[14] There are some parts of the British countryside where it is difficult to imagine any kind of major new road which would not be harmful to the landscape; there are many other areas in which properly routed and landscaped roads need not spoil the existing scenery. It seems unlikely that, in the future, road builders will be allowed to choose routes for new roads which are obviously environmentally disastrous when some better alternative is available, even if the alternative is rather more expensive. The system of publishing the proposed routes of new roads, and allowing all those who wish to do so to raise objections, means that the case for the environment does not go unheard. Environmental action groups are now well organized to object to roads which appear to be badly routed, and the holding of a public inquiry is now almost standard practice when any major new road is proposed. It is true that the government, through the Department of the Environment, has the last word on road routeing, but it is improbable that a well-founded case for an improved route would be ignored. Yet, since the judgment about which particular strip of countryside is likely to be most suited to receive a new road (judged from the point of view of its effect on the landscape) must be highly subjective, it is not surprising that there are usually some people who are dissatisfied with the result of road routing inquiries. In some inquiries, where there are two alternative routes under consideration, the concern of the objectors seems to be not so much to find the best possible route for the road, but to ensure that the route followed will be the one which will not interfere with their own particular environment. The purpose of some recent route inquiries is also a little confused, in that some objectors are more interested in opposing any new construction than they are in finding the best route for a road that is to be built somewhere. Those who wish to see Britain provided with a network of modern motorways and trunk roads have argued that the whole process of hearing objections to the proposed routes of new roads causes intolerable delay to road building. In 1969 the average time needed to investigate alternative routes for a major road and to hear objections was estimated at being one to two years.[2] However, if road building is planned far enough in advance, allowance can be made for the time needed for the measurement of public opinion on the desirability of alternative routes.

The landscaping and design of new roads and bridges has received rather less public attention than has the problem of routeing. While details of design are hardly a matter for public participation, pressure of opinion could result in the allocation of more funds to allow for effective landscaping. The proper landscaping of roads to fit into the environment is largely a matter of resources, and it would be ironic if antagonism to road building by environmental pressure groups led to the building of cheap and poorly landscaped roads. One result of the reluctance to allocate generous sums to road construction in Britain has been that existing inadequate roads are often 'improved' by piecemeal straightening and widening schemes when the building of a new road might do more to improve traffic flows and be more acceptable environmentally. The Civic Trust report *Heavy Lorries* contains pictures of some examples of the disastrous effects of minor road improvements in what were once places of beauty or architectural interest.[15] The worst cases are mostly in towns and villages where attractive urban landscapes have been spoiled or destroyed. Where new roads are to be built, then the interests of the environment demand that continuing attention should be given to problems of routeing and that resources should be made available for effective landscaping. In estimating the total costs and benefits likely to result from the construction of a road, the environmental implications both of the road itself and of the traffic flows which will be diverted to it (or which it will generate) must both be considered.

The strongest case for building new roads to improve the environment is made where by-passes and ring roads can move through streams of traffic away from the streets of towns and villages where they do great harm to that environment. The present emphasis on time-saving as the main benefit from road investment means that by-passes often show a small rate of return. Although traffic will usually move more quickly along a by-pass than through the congested main street of a small town, the by-pass road often has to take a circuitous route, and net time savings may be small. Some villages in which the environment is almost ruined by traffic are very small, and the delay to traffic may be minimal. Places where there appears to be an urgent need for a by-pass on environmental grounds often still have little hope that one will be built. This is partly a result of the inadequate sums devoted to the roads in Britain in the past, and partly a result of the investment bias just described, which puts emphasis on speeding traffic rather than on reducing environmental costs. A third factor, in some cases, is the resistance of environmental or other pressure groups to the suggested route for the by-pass. Since, in many cases, nothing could be worse than the existing route (and since traffic will certainly not just disappear), it seems difficult to justify opposition to by-pass routes which prevents or delays their building.

A study of some of the roads in Britain soon reveals not only many examples of through streams of traffic forced to travel through the centres of towns and villages which could easily be by-passed, but also numerous cases where by-passes have been built but where the environmental need was not particularly strong. One example which illustrates the curious system of priorities in building by-pass roads is to compare Farringdon in Oxfordshire with Blaby in Leicestershire. Farringdon is a small town with houses and shops lining the narrow streets, with hills leading out of it, and with both the A420 and the A417 running into the town. The traffic flows contain a high percentage of heavy lorries, and conditions during a large part of the day are intolerable. Intense public feeling has been aroused and there has been a strong demand for a by-pass for years. Blaby is a relatively modern development with the old village lying just off the A426, which is now the main street. Traffic flows through Blaby can be fairly heavy because part of A426 acts as an outer Leicester ring road and motorway link. But the road is flat, only a very short stretch is closely built up, and the traffic nuisance is not remotely comparable with that in Farringdon. There has been no special agitation for a by-pass. Yet the Blaby by-pass has been built while Farringdon still has only the promise of a by-pass (given on the day that a TV programme drew attention to Farringdon's plight). It will be a matter of years before the Farringdon by-pass is built. If there is one area in which road investment decisions appear to need reviewing it is in the building of by-passes. Local Government re-organization may give greater flexibility to local authorities in handling their traffic problems, but without standardization of basic criteria for evaluation, anomalies in the ranking of road improvement schemes are bound to arise.

The situation with new 'rural' trunk roads and motorways is rather more involved. The building of new inter-city routes will bring some environmental benefits by reducing traffic flows on built-up roads. In some cases these benefits may be very considerable, and comparable to those which can result from the construction of by-passes. But many old trunk roads which are 'replaced' by a motorway still carry a considerable volume of local traffic, and the reductions in traffic noise and other social costs may not be very great. Smaller traffic flows moving at greater speeds may be nearly as noisy as larger flows at slower speeds, and they may be even more effective in making it dangerous and difficult to cross the road. Since the new road is certain to cause some social costs, however well it is routed and landscaped, it is doubtful whether new trunk roads and motorways will yield a net benefit for the environment. The case for major new 'rural' road construction must depend upon the very important 'economic' benefits which will be gained. It seems most likely that, where the new road replaces an old slow and congested route, the 'economic' benefits will much more than outweigh any net environmental losses. This would not, of course, apply to the building of new roads for which there is little demand.

One benefit from motorways which current investment appraisal techniques may not fully consider is their long-term developmental effect. Where new roads bring about a major improvement in communications between some part of the country and a port, or a major market area, new industrial development may be encouraged which can have a profound effect on the life of the area. Just as the railways brought industrial growth in the nineteenth century, so motorways and major trunk roads can bring development today. An interesting reference to the effect of motorways has been given in an article on development in North West England. 'New motorway construction has had a particularly important impact on many of these towns, and places like Wigan, Bolton, Rochdale, Oldham, Preston, Lancaster, Warrington, St Helens and soon Bury can all be genuinely regarded as motorway towns. In addition to the pyschological fillip this provides, there is an obvious advantage to many firms in being located near to motorways, particularly those involved in distribution activities, and evidence of this is now readily available in the shape of new industrial estates etc. throughout the region.'[16] Modern road communication may also yield large social benefits by enabling a vigorous economic life to be maintained in some of the more outlying regions of Britain, such as the Scottish Highlands, the Cumbrian Coast, West Wales and South West England.

The type of new road which is clearly least likely to be justified when all economic and environmental factors are considered is the urban motorway. A high proportion of traffic in the large towns and conurbations of Britain will have both the origins and destinations of trips inside the urban area. No system of major new roads could satisfy the multitudinous needs of urban traffic flows. Where the whole of a large area is built up it is usually impossible to find routes for new roads which would not involve heavy environmental costs and probably require the destruction of houses and other property. But perhaps the strongest argument against widespread building of new urban roads is the problem of generated traffic. Road congestion is to some extent self-regulating; people are aware of road conditions and so are discouraged from driving their cars into the centres of large cities. If new roads are provided in urban areas the first result may be to expand transport demand to fill the road space provided. The 'generated' traffic comes either from people deciding to make trips which they would not have made before the new road was provided, or from people transferring from public transport (which uses up relatively little road space per person) to their own cars (which are very extravagant in using road space). One bus carrying 60 passengers may be replaced by 40 cars carrying an average of 1·5 people. The famous Buchanan report *Traffic in Towns* showed that it would be virtually impossible to provide roads to meet all the potential demand for car travel in large towns.[17] Although this 'generation of traffic' argument is undoubtedly very important it has been used somewhat rashly in some 'environmental' arguments. It applies only to a very limited extent to rural roads (see for example the M50 traffic flow figures quoted above) and it is only relevant to passenger traffic (except in a very long time period). The demand for freight traffic is largely determined by the level of activity of the economy, and will not increase suddenly because a new road has been opened.

It is likely that if a full cost benefit analysis, in which all 'economic' and environmental factors were taken into consideration, was carried out on urban motorway proposals, not many would be shown to be justified. The high cost of land acquisition, the probable net increase in environmental costs, and the difficulties in dealing with local traffic flows would all count against them. This is not at all the same thing as arguing that no new motorways should be built at all. Some conurbations, like London, do have considerable flows of through traffic and of traffic which can be concentrated on a few major entry and exit routes. There are also some environmental advantages in concentrating traffic flows on a few main routes instead of dispersing them throughout a network of unsuitable roads. The environmental costs already imposed by traffic on unimproved urban roads are, in many places, already at an almost intolerable level. As the Layfield Panel of Inquiry into the Greater London Development Plan concluded in their report 'new major roads can be satisfactorily introduced into the existing built-up area without serious adverse environmental effects' providing that great care is taken in routeing and landscaping.[18] Environmental costs can be reduced by the use of cuttings and tunnels and wide verges. Injustice to those who suffer from the new urban road can be removed by adequate compensation. The high land values which inhibit acquisition may themselves be eroded if no action is taken to reduce congestion and the social costs of traffic. It is obviously impossible to evaluate the relative price of these contradictory sets of arguments without actively carrying out detailed analysis of all proposed urban roads. It may however be tentatively suggested that while the 'Los Angeles Solution' is almost certainly not viable for British cities, it would be very surprising if it were shown that London and other large conurbations needed no modern motorway-type roads at all.

One form of road investment which has not been discussed is the modernizing and general improvement of existing trunk roads. Though small-scale road improvement in towns may have a disastrous effect on the urban landscape, as we have argued, the same is not true of rural trunk road improvements. Existing roads are often far from beautiful and it is difficult to see why any significant environmental costs need result from road widening or the construction of dual carriageways. There are certainly many 'major' trunk roads in Britain where there would be large gains in increasing traffic flows speeds and reducing accidents caused by frustration leading to risky overtaking. It seems somewhat absurd, for example, that there are long stretches of A30 (such as from Exeter to Launceston) on which it is almost impossible to overtake any slow-moving vehicles with safety.

It is difficult to generalize about such an illusive concept as 'public opinion', but it does appear that the environmental advantages brought by some new roads are now more widely recognized. The views of groups like the Council for the Protection of Rural England, who are particularly concerned about the effects of road building on the landscape, are usually well represented in *Country Life*. A recent article in that journal about the route for the London Orbital Motorway (the M25), although it was entitled 'A Choice of Disasters' and detailed numerous areas where the motorway construction might have harmful effects, nevertheless concluded that the orbital road might be an unpleasant necessity. 'A motorway or something like it has a psychological appeal that will tempt the motorist away from the shorter or more obvious route. Maybe he should not be travelling at all by car, or his goods by lorry. But as long as they are, the London Outer Orbital Route is likely to be on balance an environmental plus and not a minus.'[19] The current (1975) Minister of Transport is prepared to talk about road building with a certain amount of enthusiasm. 'I have made it quite clear that building the M25 must have very high priority, not only in terms of access to docks, which is a very important aspect of it, but also as a contribution to lessening traffic congestion in London.'[20]

Current county council TPPs contain a number of references to the environmental benefits of road construction. According to the latest Warwickshire TPP, the county council have always supported the building of motorways in their area as otherwise they themselves would have to carry out work on providing alternative routes. The same TPP contained the argument that the concentration of heavy goods vehicles on existing roads would make conditions worse for some people and that 'only the provision of local by-passes and other highway works will produce any significant environmental improvements to these and other centres in the rural community.'[11] Many county councils would like to construct by-passes for environmental reasons. Thus the Lancashire TPP describes the problem in the village of Freckleton in the Fylde District. 'Freckleton Village is thus subject to heavy traffic quite out of keeping with the village and causing environmental damage and danger to community. There is no viable alternative for the industrial traffic.' Lancashire County Council therefore gave high priority to the construction of a 'long awaited' by-pass.[21] Derbyshire County Council described the problems of the village of Hayfield in the Peak District with a daily traffic flow of 5,500 vehicles, 28 per cent being heavy lorries in 1974, and 47 accidents in the past four years. A by-pass is seen as the only solution.[22] The latest Derbyshire TPP also contains a list of ten by-passes and relief roads which, under present conditions, the county council will not be able to finance until after 1980.

An important point made in the 1976–7 Leicestershire TPP, and in those of other county councils, is that motorways will take a significant quantity, but by no means all, of the traffic away from small to medium-sized towns. Three main roads, the A47, the A447 and the A5070 meet in the centre of Hinckley in Leicestershire. When the M69 is completed it is expected that 70 per cent of the longer through trips on the A47 would transfer to the motorway. This would, however, be only about 25 per cent of the traffic on the A47 passing through the centre of Hinckley.[10]

## 4 Designated lorry routes

A final problem related to roads is whether the environmental costs of transport could be reduced by limiting the freedom of some classes of vehicle to use all parts of the road system. Local authorities have for many years been able to restrict the use of roads, but have mainly used these powers where roads were physically rather than environmentally unsuitable for heavy traffic. Heavy vehicles have been excluded from roads that were particularly narrow, or lined by overhanging buildings, and they have been prevented from using weak bridges. The Heavy Commercial Vehicles (Controls and Regulations) Act of 1973 gave local authorities wider power to exclude 'heavy' goods vehicles of over 3 tons unladen weight from some roads, and also made them responsible for preparing 'designated' routes which heavy lorries must follow. The Act also contained provisions restricting the parking of lorries. The '3 tons unladen weight' limit is quite low, and would apply to many vehicles such as refuse disposal trucks, farm lorries, and wholesale delivery vans which are clearly not in the 'juggernaut' class. In July 1974 the Department of the Environment issued a consultation paper called *Routes for*

*Heavy Lorries* which suggested the development of national lorry routes for vehicles of 24 tons gross weight and over, confined mainly to trunk roads and motorways, and asked for observations from local authorities and others. At the time of writing no firm steps have been taken to initiate these national lorry routes.

Where alternative routes are open to heavy goods traffic, and where one route is clearly preferable on environmental grounds, then it obviously makes sense to direct lorries on to that route. If the preferred route is slightly more expensive for lorries, then presumably the community will accept the small increase in transport costs. There are some high amenity areas where it is apparent that the greatest possible effort should be made to exclude heavy traffic flows. But apart from these rather obvious points it is difficult to see how the adoption of restricted or designated routes can, by itself, be used to reduce environmental costs. In very many parts of the country there are no environmentally satisfactory routes for heavy goods or any other traffic. The adoption of designated routes can only be effective if it is linked to the provision of new, environmentally satisfactory routes for traffic. Staffordshire County Council have made this point very clearly. 'It is clear that the prescription of lorry routes will provide advantages only if funds are made available to construct new lorry routes to by-pass sensitive areas.'[23] Where such alternative routes are provided it is probable that most goods vehicles, and other traffic, would use them without the compulsion of designated routes. If each small town which suffered from through flows of heavy traffic banned lorries from its streets, large areas of the country would be deprived of goods transport and their economic life brought to a halt. Some route guidance and concentration of through traffic may bring small gains in a few areas, but this purely negative approach to the problem is unlikely to bring any very significant benefits. In larger towns, where the existing road network can only just carry traffic flows, it is not clear what would happen if part of the network was closed to 'heavy' goods vehicles without any new road provision being made. Attempts by Canterbury and Exeter to exclude or restrict flows of goods traffic were overruled by the Department of the Environment because there were no other acceptable routes available. The widespread adoption of designated routes and lorry bans would undoubtedly create many problems since many exceptions would have to be made for 'local' traffic. Quarries, farms, petrol stations, shops, and other traffic generators are often situated on roads to which bans would be applied.

Nottinghamshire County Council, which appears, from its 1976–7 TPP, to be particularly environmentally conscious, nevertheless has expressed doubts about the national lorry route proposals. It is argued that three more roads would need to be added to the list of approved lorry routes in Nottinghamshire.[12] The county council also considered that 'the provision of a coarse national network providing for heavy vehicles of 24 tons or over, or over 12 metres in length, as set out in the document would do little to improve the problems brought about by heavy vehicles on roads in Nottinghamshire'. On the other hand the county council considered that some specific orders banning vehicles of 3 tons and over except for access might be beneficial.[12]

The philosophy of bans on commercial vehicles which inspired the Heavy Commercial Vehicles Act appears to over-emphasize the role of goods vehicles in causing environmental problems. As was shown in Chapters Three and Six, cars and buses also contribute to the social costs of traffic and measures which deal only with lorries may be very partial remedies. A designated route system would probably mean that many cars would transfer to the roads from which lorries were excluded and their ultimate condition might be little better than it was before the ban was imposed. One possible development of the national lorry route scheme might be to link this with an increase in permitted vehicle weights. If 38 or 40 ton gross weight vehicles were restricted to a clearly defined national lorry network then hauliers would be able to judge whether it was worth while to acquire these vehicles. In the long run the adoption of designated routes could lead to movements of industry and population and would thus have important implications for regional economic development policy.

## Summary

*1. Major reductions in the environmental costs of transport can be made where routes can be found for traffic flows which are at a distance from where people live. This applies to rail, canal and air transport, but, in Britain, road construction provides the greatest opportunities for finding routes with low environmental costs.*

*2. Road investment appraisal in Britain reflects environmental factors very imperfectly. The only environmental benefit directly measured in the 'COBA' method of trunk road investment appraisal is that of reductions in accident rates. Other environmental factors may be considered during inquiries into the proposed routes for new roads.*

*3. The payment of compensation to people who suffer as the result of new road works is to be welcomed. It could, however, introduce a bias into road investment appraisal if allowance is made for the environmental costs of new roads but not for the environmental benefits on the roads from which traffic will be removed.*

*4. Some local authorities are making interesting experiments in trying to include environmental costs and benefits in a 'points' evaluation of new road projects. In the present state of knowledge it is difficult to avoid a large subjective element in the evaluation. The 'priorities' of some county councils, such as Nottinghamshire, could lead to the use of criteria for evaluating investment in county roads which are significantly different from that used in trunk road investment appraisal.*

*5. The adverse environmental impact of new roads can be greatly reduced by landscaping and careful route planning. 'Cheap' road construction may prove to be environmentally costly.*

*6. The closing of some roads to heavy lorries does not provide an easy or cheap solution to the problem of heavy goods traffic. The adoption of designated routes for lorries will only be practicable if some existing roads are improved or if new roads are built.*

Chapter nine

# Railways and the environment

## 1 Introduction

One of the most difficult tasks for a transport economist in Britain is that of presenting an impartial analysis of any issue concerning the railways. In theory road and rail are merely two alternative techniques for providing transport services. It should be possible to compare their relative advantages and disadvantages in the same way that one could contrast the rival merits of alternative methods of manufacturing steel or sulphuric acid. But it is not so. A large number of people have an emotional commitment to the railways while another (probably much smaller) group seems actively to dislike them. 'Pro-rail' and 'anti-rail' attitudes are held with a conviction and depth of feeling which go beyond those appropriate to alternative technologies. It is no great exaggeration to say that for some people railways (like sports cars) are an end in themselves. An analysis of the reasons for the attraction excited by railways is a task for the psychologist rather than the economist. It may have something to do with their part in helping to make Britain the leading industrial power of the mid-nineteenth century. It has certainly been strengthened by a fascination for the technology of railway operation and developed by the numerous books describing the romance of the old railway companies of Britain. There is also a rather strange political attitude towards the railways whereby these products of nineteenth-century private enterprise are somehow regarded as being acceptable to those who are 'left-wing' or 'progressive' in their political thinking while road transport, and particularly road haulage, is considered to be profit-making, capitalist, and undesirable. The economist cannot, or should not, ignore popular attitudes to road or rail as being irrational or irrelevant. Public opinion and sentiment are powerful and relevant factors in determining the allocation of resources between road and rail. They are also important in considering the issue of the environmental impact of the railways.

We are not, in this study, concerned with the whole question of the future of the railways in Britain, but only with the nature of their environmental impact and with the important question of whether the total environmental costs of transport would be reduced if more use were made of rail, and less of the roads. There are arguments for retaining or expanding the rail system in Britain, such as the need to provide employment for railway workers, which have nothing to do with the environmental debate. But the railways have come to rely heavily on arguments about their contribution to the reduction of environmental costs as a means of justifying claims for government financial support. Insofar as the analysis of this chapter suggests that the greater use of rail is not a very promising means of reducing the environmental costs of transport it will no doubt be seen as an attack on the future of the railways. We can only state that the analysis is not based on a preference for any particular form of transport. We neither love lorries nor enjoy playing with trains. If the general trend of the arguments is unfavourable to rail then this is because it is the way the facts appear to be. It seems to us that the attraction of the railways already described has meant that people were very ready to be convinced of the environmental advantages of rail once the environmental problem became important. It is therefore necessary to spend rather more space on examining the case for a greater use of rail than would otherwise be the case. Although, as will be argued, it seems that the environmental benefits of rail operation may have been exaggerated, this does not mean that attitudes towards the railways are irrelevant in our assessment. Environmental costs, as has been argued earlier, are largely subjective. If people prefer the noise of a train to that of a lorry (even though both noises may register at the same level on a noise meter) then the train noise is less costly to society. But costs that only exist in our minds may change, or be changed. The problem of accidents is not discussed in this chapter as it was examined in Chapter Six.

## 2 Rail and goods traffic

Claims that traffic should be transferred from road to rail usually refer to goods traffic. The greater use of rail is seen as an easy and cheap method of taking lorries off the roads. Some of the more authoritative arguments for a road to rail transfer may be reviewed briefly. The 'Independent Commission on Transport' was made up of a distinguished group of academics (including some leading economists), theologians and people in public life. Their report *Changing Directions*[1] was in many ways a stimulating and well-balanced account of the environmental problems of transport, although, as will be argued later, their view on the role of the railways can be challenged. Their basic conclusion on freight and the railways was that 'Although the great advantages of road transport for many types of goods are undeniable, the Commission nevertheless believes that the interaction between rail and road is important and that there are some valuable gains to be made by transferring traffic from road to rail.'[1] A *Guardian* editorial written at the height of the 'anti-juggernaut' campaign stated that 'It has been suggested by the Department of the Environment that monster trucks should be confined to special routes: in which case container depots would presumably be located on a network of dual-carriageway freight routes leading to the ports. As a means of curbing existing heavy traffic the proposal has much to commend it. In fact such a network already exists. It is called British Rail.'[2] Another editorial in the *Sunday Times* argued as follows: 'But it (the motor vehicle) must at least share the job with railways. Their comparative orderliness, safety, and cleanliness cannot be wasted. . . . They (the roads) should be debited with the social costs: the cost of accidents, of congestion, of despoilation and pollution. . . . A firm and generous investment programme will enable British Rail to equip itself over the years.'[3] In 1967 the philosophy behind those parts of the 1968 Transport Act which relate to freight traffic was spelt out in a government White Paper. This stated that one of the objectives of government policy was 'to make the maximum economic use of our railways as well as our roads by promoting the transfer of all suitable traffic from congested roads on to the railways; and at the same time to make available to industry generally the full benefits of the new freightliner system'.[4] The authors of the Civic Trust report on heavy lorries argued that: 'If there is any spare freight capacity on Britain's surviving railways, as there undoubtedly is, this should be exploited to the full. One ground has already been mentioned specifically – that of human safety. Another, which is closely related to the first, is that of amenity. Railways cause less environmental disruption than road transport. Their track is segregated and not shared with pedestrians so that there can be no conflict of the kind that occurs where shopping centres are also used as through roads. As the vehicles are guided they are less prone to leave the track and damage nearby property. Any noise and vibration nuisance from railways is confined to a finite zone, whereas on the roads there is little to prevent a sudden influx of heavy lorries on a route which has hitherto carried only light traffic – with a correspondingly sudden increase in noise and vibration nuisance.'[5] A report by the Council for the Protection of Rural England was rather more extreme. 'We are unconvinced that the future of heavy long-distance travel, especially freight, lies solely, or even mainly with or on the roads. The costing systems for road and rail are different and based on different criteria. On the railways, each operation or each programme has to show a profit; on the roads profitability is shown notionally only as a theoretical saving expressed in terms of "time saved to vehicle operators", "cost of accidents", etc. The two systems are therefore not comparable. . . . We recommend that a fresh economic appraisal of rail and road systems should be carried out, using identical criteria, so far as these can be established, and including environmental factors, to arrive at a true evaluation of the situation, and to determine not only the short-term advantages to the community but also the long-term costs and benefits of each system.'[6] The most clear-cut statement of the environmental advantages comes, not surprisingly, from British Rail. The following advertisement was published under the heading 'The Superhighways of the future are already built' and was reprinted in the annual report of British Rail. '10,000 miles of rapid, uncongested, non-polluting, high capacity, load carrying steel arteries. Invest in them and we'll solve many of our transport problems economically. Ignore them and we'll destroy our greatest heritage – the natural environment.'[7]

The arguments for transferring traffic to rail as exemplified by these quotations contain two main themes which have not always been clearly separated. These are that some traffic at present carried by road could be carried more economically by rail (so that transport users would themselves benefit from the transfer) and that rail has environmental advantages. The environmental arguments were given very little consideration in the 1967 Transport of Freight White Paper, although the opening paragraph, after stating that 'The basic objectives of Government policy must be to improve the efficiency of the whole transport system. . . . Quick, convenient and economical means of moving goods are vital to industry', did add that 'At the same time wider social and economic interests must not be overlooked'.[4] The main contention of the White Paper was that there was surplus capacity on the railways and congestion on the roads so that it was 'plain commonsense' to wish to transfer traffic from road to rail and to use the 'quantity licensing' provisions which were put into the 1968 Transport Act to bring this about. The White Paper policy was also built on the belief that there had been important technological changes in the railways, mainly in the development of freightliner trains, which were then thought to be about to transform rail's competitive position for carrying freight traffic. The authors of the Civic Trust report, although mainly dealing with environmental factors, were also concerned with the over-utilization of the roads while there is surplus capacity on rail. In one paragraph in the report dealing with road congestion, it was specifically argued that 'Questions of human safety and amenity aside, the nation as a whole might save money if fuller use were made of rail freight facilities.'[5]

Now if rail could carry some freight traffic which is at present transported by lorry, not only at lower environmental costs but also at lower 'economic' costs, then the main problem would cease to exist, at least as far as the economist is concerned. If a transfer of traffic to rail would result in lower private costs per ton mile, as well as bringing net social benefits, then there would be no trade-off to be considered and it would be obvious that the traffic should be carried by rail. If this was the state of affairs, however, the question must arise of how this misallocated traffic came to be transported by road when the customer paying for it would have been better off if he had used rail. There are really only two ways of explaining how such misallocation could have come about – and both of these 'explanations' have been put forward in the past. Traders might send traffic by road which could be carried more cheaply by rail if rail rates were out of alignment with costs. In the days of the rail Standard Charges System it was argued that the existence of this scheme together with the obligation of rail to publish its rates and the provisions against 'undue preference' meant that rail was not free to compete effectively against road hauliers. The widespread use of 'exceptional rates', however, gave rail considerable freedom to vary its charges to meet road competition and it can be argued that the effects of the controls on pricing policy were not very important. In any case these controls have now been abolished and since the passing of the 1962 Transport Act the railways have been free to adopt a normal commercial pricing policy for freight traffic. There are no grounds at all today for arguing that rail charges for freight have been fixed higher than they should be in relation to costs (though, as will be shown later, it may well be argued that the opposite is true). It is perhaps worth stressing again at this point that it is only internalized or private costs which are at present under discussion – the issue of how external environmental costs and benefits might affect the rail/road cost comparison is considered later. In one important respect the railways even have an advantage over road hauliers in pricing as they have greater freedom to relate prices to marginal cost rather than average cost. Rail pricing policy is based on the principle that it is worth carrying traffic if it covers immediate costs and makes some contribution to overheads. This implies prices which are slightly above marginal cost but below average cost for some traffic (with presumably, other prices fixed above average cost where demand is sufficiently inelastic so that the railways as a whole may be in surplus). Road hauliers are less free to base prices on true marginal costs because an important part of what would normally be their fixed costs (the cost of providing the road track) is in effect paid through fuel tax, and thus made into a cost which varies directly with output. If rail really has a cost advantage in carrying any particular traffic flow, then this advantage should be reflected in rail being able to offer lower rates than road hauliers.

If the inability of rail to capture traffic for which it is better suited (in narrow economic terms) than road cannot be explained by any lack of freedom, or inefficiency, in rail pricing policy then only one explanation remains. This is that the traders who are the customers for freight traffic are not the best judges of their own interests. This second line of argument, though phrased in rather more diplomatic language, was in fact put forward in the Transport Freight White Paper. As already mentioned, it was suggested in the White Paper that the development of freightliners, and other changes, were transforming the technology of British Railways. The consumers of transport services, it was argued, would be slow to adjust, or indeed might never adjust to the improvements in rail services. 'There will often be a natural reluctance by consignors to alter their habitual arrangements for the transport of their goods, unless the advantage of the rail alternative is obvious and substantial. Inertia and habit will play their part and some consignors may not even be aware of the advantage to them of the new rail services, nor of the true economic cost of their present arrangements.'[4] A survey of the allocation of freight traffic by a sample of 125 firms in the Birmingham and Black Country area suggested that the fears of the White Paper about the inertia and anti-railway bias of consignors were not really justified.[8] Thus only five firms in the sample did not appear to have given a reasonably fair consideration to the possibility of using the freightliner service (which was just being developed at the time of the survey). The survey showed that although the technique of allocating freight traffic was by no means perfect, the transport managers of firms were neither stupid nor biased and that where rail offered advantages, either in rates or quality of service, rail would be used.

The loss of freight traffic by rail and the great growth in the carriage of goods by road has been caused neither by some peculiar anomaly in rail pricing policy nor by the ineptitude of British industry. It reflects rather the fundamental economic advantages of road for carrying most of the flows of freight traffic in Britain today. Since the main concern of this study is with environmental costs and benefits any further detailed discussion of the economics of road/rail competition would be inappropriate. But a few major points may be stressed. The existence of 'surplus capacity' on some rail routes does not necessarily mean that the capacity should be used. Rail and road are different technologies and rail marginal cost for freight haulage may exceed that of road even if road capacity is overused. There are numerous examples in industry of the existence of surplus capacity in plants using an outdated technology which is being replaced by some more efficient process. Most flows of goods traffic will need to be 'transhipped' at least once from road to rail if they are to be carried by rail for the trunk haul. This transhipment not only adds to costs, but may also adversely affect 'quality of service'. 'Quality of service' includes such factors as speed of delivery and damage and loss levels and is of great importance to many transport users. It is a major influence in determining how traders allocate their traffic flows.[8] The economic problems of rail in carrying freight traffic were illustrated by estimates made of 'direct costs' and receipts for different types of rail freight traffic for a sample week in 1961. These figures are reproduced in table 49.

**Table 49 Difference between receipts and direct costs for carriage of freight traffic according to terminal conditions, 17–23 April 1961**

| Terminal conditions | Margin (+) or (−) £ | % |
|---|---|---|
| Road–road | − 68,000 | 67·9 |
| Road–station | − 27,300 | 70·2 |
| Road–dock | − 30,300 | 67·0 |
| Road–siding | − 76,100 | 79·8 |
| Station–station | − 25,500 | 72·2 |
| Station–dock | − 7,500 | 89·6 |
| Dock–dock | − 5,600 | 41·0 |
| Station–siding | − 10,100 | 97·3 |
| Siding–dock | − 10,400 | 94·8 |
| Siding–siding | +240,300 | 132·2 |

Source: *The Reshaping of British Railways, Part I* (British Railways Board, 1963).

Although there has been some discussion about whether the 'direct' costs used in these calculations were equivalent to 'true' marginal costs there can be little doubt about the general message that even in 1961 only siding-to-siding traffic earned enough revenue to cover its operating costs.

This very condensed discussion of the economics of road/rail competition for freight traffic can be summarized by stating that road haulage has gained most of the market because it represents a superior technology. The railway management is no more to blame for this than were the manufacturers of gas mantles for losing out to the superior technology represented by the electric bulb. Railways are more efficient than *existing* road haulage facilities for carrying bulk consignments of hundreds of tons direct from coalmines to power stations or iron ore quarries to blast furnaces. But most freight in Britain is sent out in small consignments for relatively short hauls and haulage by road is, at present, the most efficient technique for carrying this kind of traffic.

There are, therefore, no grounds for supposing that there is any significant misallocation of traffic between road and rail if this is to be judged on narrow economic grounds. If there is a case for any large-scale transfer of traffic to rail then it must be based on showing not only that there would be environmental benefits but that these would be large enough to outweigh the 'economic' losses which would undoubtedly result. What then are the environmental advantages of rail compared with road transport? This question cannot be discussed sensibly on a global 'road' versus 'rail' basis but only by comparing particular situations where road and rail are alternatives to each other.

The most objectionable type of road freight traffic is probably through traffic in small towns and villages and in urban shopping centres. The Civic Trust report, for example, made the point that with railways 'there can be no conflict of the kind that occurs where shopping centres are also used as through roads'.[5] Now if there was an alternative rail route available to carry a particular flow of traffic all the way from its origin to its final destination which was at present being carried by lorry through a shopping centre or a small town, it would obviously make sense to consider using the rail service. But this simple situation is rarely, if ever, found in the real world. In many cases there may be no rail route at all to cater for the stream of traffic concerned, or it may be so roundabout that the costs of using it would be quite prohibitive.

Traffic carried by road through a particular town centre does not, of course, all originate from one consignor nor is it destined for a single consignee. Traffic originating from a wide area which is being conveyed to a large number of different ultimate destinations may be funnelled along a particular through road route. It is likely to be very much more difficult to concentrate this traffic on a single rail route with its very limited number of access and egress points, and with two transhipments being necessary, than it is to send it by road. With the development of the motorway and trunk road network the kind of road traffic flow passing through small towns is likely to be cross-country traffic flows, or short hauls, for which the rail alternative is particularly unsuitable or nonexistent. As argued in Chapter Eight, it is highly probable that the best solution to the problem of through traffic in small towns and shopping streets is the construction of a road by-pass. The railway alternative, if available at all, will usually be so expensive (especially when it is remembered that, without a by-pass, car traffic would still go through the towns) that it can be dismissed. Rail can offer little environmental advantage for traffic terminating in small towns since it will have to be carried by road from the rail goods depot, and the total mileage on the town roads is unlikely to be decreased significantly and may even be increased. The situation may be illustrated from a simple model of a town as shown in figure 4. Suppose that the flow of terminating traffic

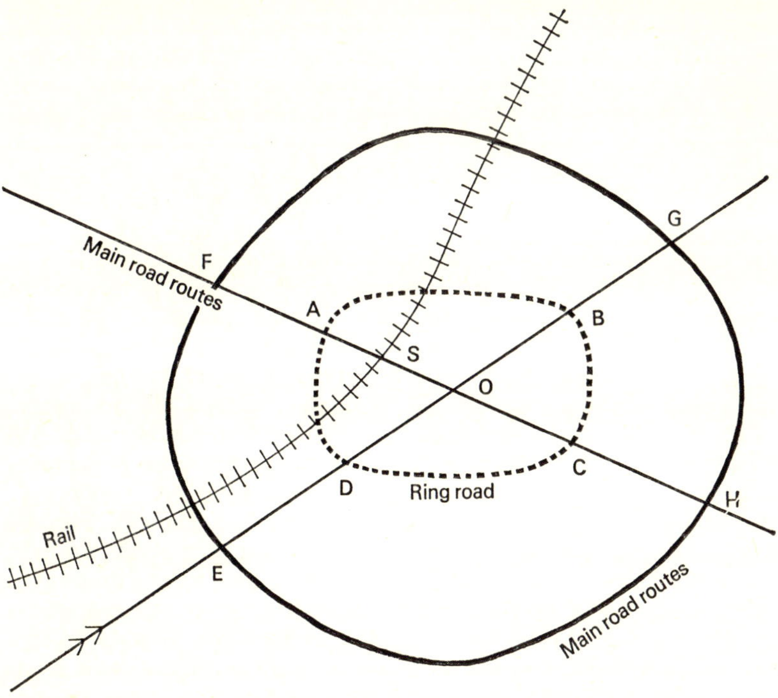


*Figure 4. Possible reduction in lorry road mileage from using road and rail.*

for the town, which can be called Modelton, all came from the South East either along the arrowed road entering at E or along the rail line. The circle E F G H represents the built-up area of Modelton and it is assumed that there are four factories or warehouses receiving incoming traffic at A, B, C, D which are equidistant from the town centre O, and from the edge of the built-up area E F G H. Now, if it is assumed that there were no suitable ring roads linking the four traffic receiving areas, and that one lorry could carry the total daily inwards traffic for all four receiving sites, then the distances run on roads inside Modelton if road or rail transport were used, can be compared. Using road transport entirely and letting m represent the road mileage from E to D, and n that from D to O, then the total distance run on roads inside Modelton would be equal to $8n + 2m$ (since $OA=OB=OC=OD$). The distance travelled by a single road vehicle operating from the rail goods depot S would be equal to 8n. The reduction in lorry road mileage if the consignments were carried by rail to S would therefore be equal to 2m. If x lorries were required to carry the traffic then the reduced road running mileage resulting from using rail would be 2xm. The net environmental benefits from using rail would depend upon the nature of the road system in Modelton. If ED, on the outskirts of the town, was a dual carriageway separated from residential development then the benefits from using rail would be negligible. It is generally true that the social costs of lorries will be higher on roads nearer the town centre and the likelihood is, in a small town, that the lorry mileage over inner, congested roads will vary very little whether delivery is made entirely by road or via a rail goods depot. For the distribution of goods in the town centre area around O the same arguments apply. The mileages run over the actual shopping streets would be similar by either mode, the only saving from using rail being in avoiding the road run-in from E. Against the very small benefits from using rail would have to be set the high 'economic' costs, costs which represent resources which could be used to improve the environmental aspects of Modelton's road system. As will be argued more fully below, the rail operation itself is also not entirely without environmental cost. If any significant volume of traffic were involved an extra train would have to be run into S, and the rail route might well pass through a built-up area, so that the operation of the train would result in some disbenefits of noise and air pollution. It is quite easy to find situations in which a town has a goods station near the congested centre, and its main industrial area (which generates inward and outward traffic flows) on the outskirts, so that use of rail to the goods station and delivery by road from there would involve higher environmental costs than using road haulage exclusively. In the Modelton example this would clearly be the case if the main traffic receiving area was centred at D. Rail environmental costs would also exceed those involved in using road, in the Modelton case, if SA was highly congested but there was a good ring road (which could carry vehicles without any significant environmental costs) connecting the four sites at A, B, C and D.

The area (apart from bulk mineral traffic) in which rail 'economic' costs are likely to be most competitive with road costs (subject to qualifications about the future of rail discussed below) is in carrying large traffic flows between major destinations. But it is also for this kind of traffic that the environmental advantages of a transfer to rail are least obvious. Where major inter-city traffic flows still pass through small towns the remedy of building a by-pass is likely to be preferable to a rail transfer. Rail would not normally have the physical capacity to carry more than a small proportion of a large through traffic flow, and the removal of a proportion of lorries, while many other lorries and all the private cars continued to travel through the centre of the town, would not bring any great environmental advantage. Where the inter-city traffic flow is carried by motorway or by dual-carriageway trunk road then it is not clear that there would be any net environmental advantages in transferring traffic to rail. No doubt some motorists would like to see fewer lorries in the slow lanes of motorways, but in other ways carrying traffic by lorry on motorways has a lower environmental cost than sending it by rail. The Council for the Protection of Rural England's sentence in praise of rail, 'Their track is segregated and not shared with pedestrians so that there can be no conflict of the kind that occurs where shopping centres are also used as through roads' applies equally to motorways. In addition inter-urban motorways are segregated from homes and built-up areas much more effectively than railways, which usually pass through the centre of every town on their route. The M1 from Leicester to London, for example, does not go through many urban areas (with the possible exception of the short stretch between Luton and Dunstable), whereas the Leicester–London rail route passes through or near to the centres of Leicester itself, Market Harborough, Kettering, Wellingborough, Bedford, Luton, and St Albans, as well as those of many smaller towns. If an efficient inter-urban rail system with surplus capacity parallels a road route which is becoming overloaded then, of course, it makes sense to use the rail route for some traffic. But rail routes are so used already and any large-scale transfer to rail on environmental grounds is neither possible nor desirable.

The remaining types of traffic flow which have not been discussed are traffic entering or leaving very large conurbations, and dock traffic. Large conurbations may differ from small towns like 'Modelton' or even from large 'isolated' cities (like Leicester or Bristol) in possessing a network of urban rail services. This means that there is at least the possibility of using intra-urban rail routes to carry consignments close to their ultimate destination (although the final delivery will have to be made by road) and thus reducing lorry road mileage on congested urban roads significantly. The greatest opportunity for the use of intra-urban rail routes would appear to be in our largest conurbation, London. Traffic coming into London can sometimes be sent to different terminal goods stations according to its ultimate destination. It can then be moved around the rail system in inner London from, for example, Camden in the north, or Willesden in the west, to goods depots like Bricklayers Arms in the south. This seems to be the one area where the social benefits of carriage by rail are likely to be high enough to make some interference with the market mechanism, presumably in the form of a rail subsidy, worth considering. The 'economic' costs of intra-urban rail working are likely to be fairly high because of the amount of transhipment and the very short hauls involved.

The main 'economic' advantage for rail in carrying dock traffic is that this will normally be containerized for its sea journey so that it would appear to be very suitable traffic for carriage by freightliner. The recent tendency, however, has been for container depots to be established some way from the docks and to be served by road. The main point as far as environmental benefit is concerned, however, is that very many docks now have no direct rail link, or no link which can be conveniently used. The *Guardian*'s suggestion that British Rail is equivalent to a 'network of dual-carriageways' leading to the ports is a great oversimplification. The need to transfer dock traffic to road for the last part of its journey means that the railways have no particular environmental advantages here.

In order to illustrate some of the points made about rail as a superior environmental transport alternative to road a brief study was made of some of the environmental implications of sending traffic to various destinations from Leicester and district by either road or rail. These practical examples are listed and described below. They all relate (except where otherwise specified) to consignments in the 1–3-ton weight range.

## Leicester to Millwall docks

*Rail*
Collected from factory in Leicester by covered van or flat lorry. Taken to goods depot in Leicester (Humberstone Road/ Northampton Street) which is very near to the city centre. Train made up in Leicester for St Pancras goods station. Train taken to adjacent Churchyard sidings and rail-van (with dock consignment) tripped and taken to Camden goods station. Consignment carried from Camden to Millwall by road.

*Road*
More alternatives possible than by rail. Could be carried by motorway and dual carriageway from outskirts of Leicester to within about 3 miles of rail terminal point in Camden. More likely to leave motorway in North London and use urban (but not Central London) roads to docks.

*Environmental comparison*

Carriage by rail could bring a loss in Leicester because all traffic would need to be taken by road to the goods station near the city centre. Complete carriage by road would mean that traffic from outer industrial areas could avoid the city centre. Carriage by rail would mean a slightly reduced lorry road mileage in London, but the final road haul would be through some of the most congested areas of inner London. It is doubtful whether carriage by rail would bring any net environmental benefits.

## Leicester to Croydon

*Rail*
Consignment would be sent via Whitemoor (near March in Cambridgeshire) to Temple Mills (near Stratford in East London) and then round inner London rail lines via Willesden and Clapham Junction to the nearest suitable South London depot, which would probably be Norwood, and thence to its destination.

*Road*
As before to some point in North London, then either through Central London or via the South Circular route.

*Environmental comparison*
As before for Leicester end of haul. Carriage by rail would clearly mean a very considerable reduction in lorry mileage on congested London roads. But railway operation in London also involves environmental costs through noise, pollution and 'visual intrusion'. The 'economic' cost of rail with the roundabout route from Leicester, the marshalling at Whitemoor and Temple Mills, and the meandering operation round the inner London rail routes must be very much higher than the equivalent road 'economic' costs.

## North London to Loughborough

*Rail*
By road to Camden, rail to Leicester goods depot, and then road to Loughborough.

*Road*
Motorway to Exit 23, short all-purpose road into Loughborough.

*Environmental comparison*
Rail carriage would involve considerable environmental disadvantages at both ends of the haul. It is likely that the lorry mileage on the more congested inner London roads would be greater (since the motorway has access points on the outer part of the built-up area, while Camden is in inner London). Delivery by road from central Leicester, through the industrialized and heavily trafficked areas of North Leicester and along the largely built-up A6, clearly involves higher environmental costs than using the motorway to the outskirts of Loughborough.

The contention, which is illustrated by these case studies, that the greater use of rail for freight transport would do little to improve the urban environment, is not new. The Geddes Committee on commercial vehicle licensing argued, in their report, that there was no case for protecting railways in order to relieve urban congestion since the railways were not generally available to replace road routes in urban areas.[12] In a memorandum to the Expenditure Committee's Sub-Committee investigating the problems of Urban Transport Planning, the Department of the Environment argued that a large proportion of freight traffic carried in a large urban area starts and completes its journey inside the conurbation. 'The bulk of goods vehicles journeys in urban areas involve collection or delivery at premises within the area. In the case of the larger cities the bulk of journeys are made entirely within the urban area; even for heavy vehicles (over 3 tons unladen weight) the average journey length during 1971 was only some 30 miles.'[13] The same memorandum discussed the possibility of transferring traffic to rail. 'Little alleviation to the problems of goods transport in towns can be found by transfer to rail or water transport. These are mainly suitable for bulk movements between a limited number of terminals. . . . In the main urban centres, where most of the goods would still have to be collected and delivered from the rail yards or wharves, there would be little or no environmental benefits.'[13]

A fundamental reason why transfer to rail cannot remove the problem of the environmental costs of road transport is that rail does not have the capacity to carry more than a small part of road freight. The record of the last twenty years is not so much one of a transfer of traffic from rail to road but rather of an increasing total demand for freight traffic, all of the increase being in road-borne traffic. Thus in 1954, before the Beeching cuts, the combined road/rail total was 41·6 thousand million ton miles (rail 22·1, road 19·5), whereas the equivalent total in 1974 was 69·8 thousand million ton miles (rail, including passenger parcels 14·8, road 55·0).[10] Even in 1912, rail carried only 513·6 million tons although it had a monopoly of long-distance (inland) haulage, 20,038 miles of line open for traffic, and the figures included a large tonnage of coal traffic.[11] In 1975 road and rail together carried an estimated 1,881 million tons of goods traffic.[11]

At the end of 1974 British Rail had 11,289 route miles of track open to carry traffic.[14] In 1973 (1 April) there were 212,034 miles of public roads in Great Britain, of which 29,799 were trunk roads and principal roads.[15] The ratio of road miles to rail route open was thus rather more than 18 : 1. But this figure, of course, does not reveal the full limitations of the rail service. The total number of stations receiving both passenger and freight traffic open on the British Rail system at the end of 1974 was 93, and another 435 stations were open for freight traffic only. There were thus only 528 points of access for general merchandise traffic for the rail system for the whole of Great Britain. (There are also a number of private sidings connecting with rail but these now carry bulk traffic almost exclusively. It is generally hopelessly uneconomic to use pick-up trains and private sidings to collect or deliver consignments in the weight range, up to 22 tons, of traffic at present carried by road vehicles). Except for motorways, every road is open to direct access from premises built alongside it. Almost every factory, warehouse, shop and house in Britain has immediate access to the road system. Every factory of any size has, in effect, its own private road 'sidings' which may link directly with assembly lines and with receiving stores. The representatives of some firms in the Birmingham-Black Country area said that they regarded their road delivery and inter-factory service as 'a continuation of the assembly line'.[8] To provide traffic for the 528 rail general goods access points there were 90,000 manufacturing 'establishments' in Great Britain in 1968,[16] and (in 1971) 485,000 shops (not counting service establishments) needing to be supplied with consumer goods.[17] Besides shops there were 19·0 million houses and flats (Great Britain 1972) and an unknown number of warehouses, depots, docks, building sites and quarries receiving and sending out freight traffic, as well as numerous inter-factory flows of traffic. The immediate linking of virtually all of these generators and receivers of traffic flows to the road system, and the absence of any direct rail connection for the vast majority of them, is a fundamental fact which must have a profound effect on the allocation of goods traffic between the two modes.

To determine how much traffic could be transferred from road to rail would involve carrying out a difficult and extensive survey. The physical capacity of rail must obviously be related to the demand for freight transport. Rail surplus capacity on the London to Penzance route is not relevant to freight flows between Birmingham and Liverpool. In the absence of detailed knowledge, estimates of rail capacity tend to be influenced by the viewpoint of the estimator. In their memorandum to the Expenditure Sub-Committee on the problems of Urban Transport Planning, the Department of the Environment argued that 'even if the freight tonnage carried by the rail system as a whole could conceivably be increased by half, the decrease in road traffic would be only between 1 and 2 per cent'.[13] This underestimates the possible environmental improvement on the roads from the transfer by relating the fall in vehicle miles to all traffic rather than to freight traffic alone (though this procedure may be more justifiable if the concern is with road congestion rather than general environmental costs). The *Changing Directions* report goes to the other extreme by assuming that the average length of haul of the transferred traffic would be 124·9 miles (compared with the actual average rail haul in 1974 of 79·5 miles).[14] This would represent about 22 per cent of 1974 road ton mileage. But there is no evidence that (as the 'Independent Commission's' analysis assumes) anything like 60 per cent of road haulage tonnage travelling over 150 miles could be transferred to rail. An increase of 12,490 million ton miles carried would involve a near-doubling of the work done by the railways, not an increase of 50 per cent and the 50 per cent figure is itself hypothetical.

The roundabout nature of the rail routes between many places is now a serious problem. A traffic flow being carried by rail at the time of the *Living with the Lorry* study, which appears to be ideal rail traffic, was a flow of oil tanker wagons from the refinery at Fawley, near Southampton.

to Colwick, near Nottingham. But there is now no kind of direct rail link between Southampton and the east Midlands, and even train-load traffic like this flow of oil tankers must follow a roundabout route which must partly offset the 'economic', if not the environmental advantages of sending this traffic by rail. Another flow of traffic being carried partly by rail (details of which were given to us during the research carried out for this study) is of 600 tons of chemicals from the Runcorn area to London and the South East. This is now carried by rail to Willesden where the liquids are pumped into tank 'farms' and both liquids and solids (which are palletized) are eventually distributed from there by road. Since virtually the whole of this trunk run could now be made by motorway, and since the Willesden distribution terminal is situated very near to the centre of the London conurbation, the environmental advantages of this arrangement are not particularly obvious. If it were not worth while using rail for this traffic flow on 'economic' grounds then it would hardly be justified for environmental reasons.

There is also another problem, illustrated by the Leicester–Croydon flow in the case studies, which is likely to arise when the increased use of intra-urban rail lines for carrying freight traffic is under consideration. This is that these lines may already be fully used for passenger traffic. This obviously applies particularly to the Southern system in South London and the large flow of passenger traffic into Central London is one of the reasons for the need to use a circuitous route to transfer freight from goods stations like Temple Mills to South London. In other words the rail 'surplus capacity' may not exist where its use could yield the greatest environmental benefits.

It is thus quite misleading to list the environmental advantages of rail – 'their comparative orderliness, safety, and cleanliness',[3] their lower 'environmental disruption',[5] their 'uncongested, non-polluting high capacity'[7] – as if these benefits could be gained by any transfer of traffic from road to rail, or as if the choice was between door-to-door road or rail transport. The most undisputed 'environmental' advantage of rail is, perhaps, its greater safety. But this does not mean that any particular road-to-rail transfer would make any significant contribution to reducing accidents, since most accidents occur on urban roads, and as has been shown, a 'road-to-rail' transfer may actually increase the mileage run by lorries on urban roads. It is indeed far from certain whether the removal of a percentage of lorries from trunk roads and motorways would reduce accident rates at all.

Even where it is possible to substitute an all-road haul by an all-rail haul (rather than by a road-rail-road haul) the environmental advantages of rail operation are not always so apparent as some environmentalist groups assume. Some discussions of road-to-rail traffic transfers appear to be based on the belief that the railways do not impose any environmental costs on the community. But railways are neither silent nor invisible. They divide communities, and the emissions from diesel-engined locomotives are not very different from those of diesel-engined lorries. There may be environmental benefits if a very small number of trains can do the work of a large fleet of lorries, or if a rail route can be used which is further away from people and houses than the road route which it replaces. But neither of these circumstances are a necessary property of all road-to-rail transfers. The comparatively generous attitude adopted towards rail environmental costs can be illustrated by considering the treatment of the noise issue in the 'Independent Commission's' report. The report recognizes that road traffic and aircraft noise is likely to increase as traffic increases but concludes that 'Railway noise is not likely to rise above its present level'.[1] But if the recommendations of the Commission were followed and traffic was transferred to rail, then more trains would be run and more noise would be created. A recent study of rail noise contained the argument that 'railway noise must be considered as a growing environmental disturbance'.[18] The references in this paper were mainly to high speed passenger trains, but a freight train travelling at 100 kmph can have a noise peak of 85 dB(A) at 50 metres from the track.[19] Speed has an important influence on rail noise and the 'Independent Commission's' report is distinctly optimistic in believing that the operation of the Advanced Passenger Train will not create noise problems. As was suggested in Chapter Three, people in Britain appear to be more tolerant of railway noise than they are of road traffic or aircraft noise. According to Walker's estimates the noise levels for 100 metres on each side of a rail track are higher than the 68 dB(A) $L_{10}$ value recommended by the Department of the Environment as the maximum acceptable level for road traffic noise.[19] It is interesting to speculate about the reasons for the comparatively uncomplaining acceptance of rail noise in Britain. It may be that our techniques for measuring noise nuisance are imperfect and that there is some quality in railway noise, undetected by noise meters, which makes it more acceptable to the human ear than road or air noise. A more likely explanation is that rail noise has fallen with the reduction in the scale of railway activities The railways have been producing noise for a much longer period than aircraft or motorized road traffic. People may have become accustomed to rail noise in a way which reduces its nuisance value. They have also had time, and the opportunity (since rail noise in Britain is more localized than road traffic noise) to move away from affected areas. If railway noise and other rail environmental costs such as visual intrusion are reflected in lower house prices those people who are less troubled by rail noise (or whose low incomes lead them to put a low money value on freedom from rail noise), may be induced to move into the areas which are near to railway lines. But, as Walker points out, the acceptance of existing rail noise does not mean that an increase in rail noise (whether caused by changes in technology, heavier rail traffic, or the opening of new rail lines) will be without cost to the community. The opening of an extension to the high speed Shinkansen rail network in Japan in 1972 led to widespread protests about the noise, and to the preparation by Japanese National Railways of plans for reducing noise over the whole high speed network.[19] In Britain the proposal to build a new rail line to link London with the Channel Tunnel led to many protests about the expected damage to the environment.

There are other possible explanations of the different attitudes towards road and rail noise which lie in the area of psychology and sociology rather than of economics. The general dislike for lorries and the publicity given to the 'anti-juggernaut' campaign may have made people more aware of the noise created by lorries and other forms of road traffic. It is an interesting possibility that social costs which reflect subjective valuations of disbenefits may be increased or reduced as a result of publicity campaigns. It may also be the case that the people who are most exposed to rail noise, such as those who live in houses adjacent to rail lines in inner city areas, are those who are least likely to make effective protests.

The idea that railways were not harmful to the environment, or could solve the problem of goods transportation inside urban areas, would have seemed very curious to those who lived in the 'Railway Age'. A major study by John R. Kellett, *The Impact of Railways on Victorian Cities*, gives innumerable instances of the devastating environmental effects of the building of the railways.[20] Victorian environmentalists who complained of railways placing 'an impassable and hideous wall between the most important populous parts' of urban areas would be amazed at the present belief in the railways as being somehow environmentally pure. It is also interesting that even when railways were at the peak of their importance they were not able to deal satisfactorily with the carriage of traffic inside towns. According to Kellett the small miscellaneous loads of Birmingham or London manufacturers made the railways slow and uneconomic for the local movement of goods. Within a ten-mile radius of central London by far the greater part of retail distribution remained in the hands of carters. In 1900 vans and carts carried 90 per cent or more of merchandise and shop goods over a two-hour journey or ten-mile radius from central London. The concentration of long-distance traffic arriving or being despatched from the railway stations and termini also led to considerable congestion. Congestion around London termini was such that one could 'walk over the roofs of vans and buses as readily as over the united up-raised shields of the Roman soldiers outside the walls of some beleaguered city'.[20]

### 3 Rail and passenger traffic

The possibility of gaining environmental benefits by transferring passengers from road to rail has received much less attention than have proposals to transfer goods traffic. A major reason for this has been a greater concern about the environmental costs of goods vehicles than about those caused by cars. It has also been recognized that transferring passengers from cars to buses and coaches will reduce environmental costs, and that this policy may be easier to carry out than a car-to-rail transfer. Another factor which has not been explicitly spelt out, but which is certainly not without significance, is that it is apparent that people can choose rail travel if they wish. The image of private profit-maximizing traders who send their goods by road regardless of the social costs involved may be made to appear convincing, whether or not it accords with the facts. But if the majority of people in the community choose to travel by road rather than by rail, and to bear the costs of greater accident risk, noise and pollution, then it is less easy to argue that they should be compelled or persuaded to use the railways. (There are, of course, environmental costs which car travellers impose on pedestrians and people living and working near roads, but many of these people will themselves be car travellers on some occasions.)

If the pressure to transfer traffic from road to rail is less for passenger than for freight traffic, the possibilities of organizing a transfer are probably greater. The environmental advantages of rail, in so far as they exist, are mainly related to the high carrying capacity of trains. A train carrying 800 tons of freight or 500 passengers is likely to cause less environmental damage than the equivalent 40 lorries or 125 cars. But, as has been argued, it is very difficult to organize full train loads of general freight traffic. It is much easier to concentrate passenger traffic for 'mass' carriage because people can transport themselves to railway stations. Passenger traffic does not have the transhipment problem of freight traffic (though there may be a problem of finding parking space near stations when a transfer of mode is necessary). For the journey to work, rail has the great advantage that rail terminals may be within walking distance of the central business districts of many towns.

## 4 Rail and fuel consumption

The 'fuel problem' is not strictly an environmental issue. There is no externality problem since the cost of producing fuel is fully represented in the price. In Britain fuel prices for road vehicles, but not for the railways, are greatly increased by taxation. This tax element is probably high enough to allow for the scarcity value of foreign exchange which must be used to buy petroleum products. Whether fuel prices are high enough to safeguard the interests of future generations is an issue which is highly dependent on value judgments, and one which we will not attempt to determine. But although fuel is just one input required by the transport industry, whose scarcity can be reflected in its price (in the same way as that of other inputs such as manpower, rubber and steel) the possibility that supplies of petroleum based fuels will be exhausted during the next hundred years does have great significance for the transport industry. It seemed therefore that some brief consideration of the fuel problem should be included in this study even if it is not a true 'environmental' issue. The discussion is included in this chapter as if is sometimes claimed that the railways make a more economical use of fuel than does road transport.

The importance of transport as a consumer of energy varies with the measurement used. Measured in terms of tons of petroleum equivalent, transport consumed about 14 per cent of total energy consumption in the EEC in 1973 but about 25 per cent of petroleum fuels.[21] Transport in the U.K. consumed about 27 per cent of the tonnage of petroleum fuels in 1972.[22] But measured on a 'heat supplied' basis transport energy consumption amounted to 21·2 per cent of all energy supplied from petroleum in the U.K. in 1974.[23] Cars and taxis consumed about 13 million tons of fuel in the U.K. in 1972, petrol lorries 2·5 million tons, diesel-engined lorries 5 million tons, the railways 1 million tons and aircraft 4 million tons (these figures do not include oil used to generate electricity for electrified rail routes).

There are some circumstances in which it is obvious that rail would use fuel more economically than road transport. A train hauling 1,000 tons of mineral traffic from A to B would use much less fuel than 50 lorries carrying the same load between the same two points. Similarly a passenger train carrying 200 passengers will make a much more economical use of fuel than 100 cars running between the same two points. But the extent to which rail has an advantage in fuel consumption per ton mile or per passenger mile in actual operating conditions is much more uncertain. One study which attempted to measure the energy consumed in relation to output for different transport modes was carried out by the Netherlands Road Transport Institute. The results for urban passenger transport per passenger/km (bus consumption = 100) were:[24]

| | |
|---|---|
| Moped | 75 |
| Bus | 100 |
| Underground railway | 150 |
| 'European' cars | 300 |
| 'American' cars | 400 |

For 'trunk' passenger transport the estimated figures were:[24]

| | |
|---|---|
| Bus | 100 |
| Train | 150 |
| Car | 300 |
| Aircraft | 600 |

The estimated figures for energy consumption for goods transport, with rail as 100, were:[24]

| | |
|---|---|
| Pipeline | 60 |
| Water | 57 |
| Rail | 100 |
| Trunk roads | 300 |
| Aircraft | 4,000 |

These figures suggest that for passenger transport rail has a fuel consumption advantage over cars but not over buses, and that it is more efficient than lorries in using fuel for goods transport. But it is doubtful whether figures at this level of aggregation are very meaningful. As Bayliss has pointed out,[21] ton miles and passenger miles are both so heterogeneous that aggregation for any kind of productivity measurement is illegitimate. The output of aircraft cannot be compared with that of other modes which may take 20, or even 50 times longer to deliver people or goods. Load factors are crucial in influencing intermodal energy consumption comparisons. While fully-loaded goods trains will have a much lower fuel consumption per ton mile than lorries, trains with the kind of load factors which are usually achieved for carrying general goods traffic are much less favourably placed. It is also apparent that comparable energy consumption measurements should relate to total hauls on passenger journeys. Professor Cheeseman of Southampton University made some estimates of the relative energy consumption of different transport modes in a paper originally presented to the Universities Transport Studies Group Conference at University

College London in 1973, and quoted in Gwilliam and Mackie.[25] The original figures, which related to a mixed goods train and a 30-ton truck (loaded to 60 per cent of capacity), actually suggested that the lorry was more efficient, yielding 100 ton miles of 'output' per therm compared with only 66·7 ton miles for the train. Cheeseman has now reworked these figures, however, and concluded that a mixed goods train is 1·2 times as efficient as a 30-ton lorry in using energy, and a modern freightliner about twice as efficient for comparable trunk hauls.[26] It is interesting to note, incidentally, that Cheeseman's calculations also dealt with land use, and showed that rail was more wasteful than buses in using land for passenger transportation. According to Cheeseman rail produced 10,100 and bus 28,900 passenger miles per hour per acre (assuming 60 per cent load factors in both cases).[25]

It is apparent that comparative energy consumption measurements should relate to total hauls or passenger journeys rather than to the trunk journey alone. Thus in comparing road and rail energy consumption for goods traffic, allowance must be made for the terminal collection and delivery by road which is necessary for most general merchandise traffic. Trunk hauls themselves may be of different lengths by different modes. A French study showed that the average road distance (for the trunk journey only) was only 88 per cent of that of the equivalent rail distance.[27]

There is no case for any direct action to alter the present distribution of traffic between road and rail in order to reduce fuel consumption. Fuel is only one scarce resource; shortages of other resources such as land, zinc and steel may become of critical importance in the future. There is no reason why we should not continue to rely on the price mechanism to ensure an economic use of fuel. If fuel became so expensive that a rail advantage in fuel efficiency outweighed all the other factors favouring the use of roads, then road transport would become more expensive than the equivalent rail transport and traffic would move back from road to rail. Increasing fuel prices will tend to reinforce a rail advantage where it already exists; that is in the carriage of bulk mineral traffic from siding to siding. They are unlikely to affect the modal split of other goods traffic, or the distribution of passenger traffic between bus and rail. Road haulage fuel efficiency is artificially reduced at present by the maximum gross vehicle weights imposed in Britain. If the maximum permitted gross weights of lorries was raised to 38 or 40 tonnes, average fuel consumption figures would be improved.

The argument that it is not necessary for the state to interfere with the allocation of traffic between road and rail in order to conserve fuel supplies does not mean that the probability of the exhaustion of petroleum-based fuel supplies during the next hundred years can be ignored. Considerable further price increases (brought about through taxation) may be considered necessary in the future in order to restrict the demand for petroleum fuels. The greatest scope for economies is in private motoring. Speeds can be reduced, load factors increased and more low-consumption cars can be substituted for high-consumption vehicles. It is difficult to find any rational case for using cars with a petrol consumption of 10 mpg when (given the current speed limits in Britain) journey times will by very little different from those which can be achieved in a car consuming 30 mpg. When petroleum fuels are eventually exhausted then the internal combustion engine must be replaced, probably by some form of electric motor. If battery operation is still not developed by that time then it may be necessary to provide overhead electricity supply. There is no obvious reason why this should involve running on steel rails rather than on a road surface.

## 5 The costs of rail operation

It has been argued so far in this chapter that there are some circumstances in which transport by rail may involve lower environmental costs than road transport, though the rail advantage is not so great or universal as has often been claimed and the opportunities for transferring additional traffic to rail are strictly limited. If we assume that there are some environmental benefits to the community resulting from the *current* use of the railways for carrying passengers and goods, the important question arises of how much the community has to pay in order to obtain these benefits. It is possible that the costs may exceed the value of the benefits, or that equal environmental benefits could be obtained by following some alternative and less costly policy.

The cost of operating the railways so that they can carry existing flows of goods traffic and passengers is met in two ways. The bulk of the cost is recovered in the same way as it is in other industries by making charges to the consumers of the services which the railways provide. But the railways have not earned enough revenue to cover all their costs since 1952 and have been forced to rely on government subsidies. British Rail accounts do not set out their subsidy position with any great clarity. Grants received under the 1968 Transport Act for running 'unremunerative' passenger services are treated as revenue and added into the total of net receipts. This leads to some confusion about the actual size of the losses incurred by the railways, since the figure generally reported in the media relates to the loss after some subsidies have been received. Even Gwilliam and Mackie[25] write of a rail 'overall surplus' in 1969 and 1970, though the surplus was only achieved after payment of substantial subsidies. The Board's 'overall working surplus' in 1970, for example, was £9·5 million. But this includes 'revenue' of £61·7 million from grants paid under the 1968 Act.[82] Before 1969 the government simply met the total rail loss; from 1969 onwards grants were paid and introduced in a different place in the accounts. The dramatic change from a loss of £147·4 million in 1968 to a 'surplus' of £14·7 million in 1969 was merely the result of a change in the method of accounting (and the writing-off of a large slab of rail debt plus the removal of the loss-making sundries business to the National Freight Corporation's accounts). The belief of the Railways Board that they somehow 'earned' the government grant of 1969 in a way in which they did not earn the government support of previous years is not tenable. It is also something of a myth to suppose that the railways after 1969 ran 'unremunerative passenger services' for the government's benefit which, if they had closed them down, would have reduced the rail deficit by the amount of the grant. If all grant-aided passenger services had been withdrawn, very few services would have been left and freight traffic would certainly have been unable to bear all the overhead costs formerly shared with passenger traffic. The 1970 Railways Board Annual Report contained the interesting argument that 'The grant to aid specific services is not a charity but is simply a method of providing the community with services which the appropriate authority has ascertained are required. It is not a subsidy to British Rail but to the travellers using these services; and it is akin to the transfer payments and subsidies arising from other social services provided by Government.' Despite this curious argument all government payments to the railways will be described as subsidies, whether they were paid before or after arriving at the balance of the Profit and Loss account.

In 1974 the Railways Board made a loss on the Profit and Loss account (including non-rail activities) of £312·1 million. In British Rail accounting this was made up of a loss of £157·8 million plus a grant of £154·3 million.[29] But besides this grant, other grants of £23·5 million were received on the Capital Account (£22·7 m. under Section 56 of the 1968 Act and £0·8 m. for port modernization grants), and the state also had to make good the rail deficit and 'meet the Boards cash flow shortfall' which cost a total of £215 million. The total subsidy situation in 1974 was therefore:

| | £ million |
|---|---|
| Grant for unremunerative passenger services | 154·3 |
| Section 56 and rail modernization | 22·7 |
| Cash flow shortfall | 215·0 |
| | 392·0 |

The government has never listed the social benefits it considers the community gains from the subsidies paid to the railways. Under the Railways Act of 1974 even the attempt to identify the grant for loss-making passenger lines has been abandoned. This Act extinguishes yet more railway debt and empowers the Secretary of State for the Environment to pay 'compensation' to the railways of up to £1,500 million in total.[30] It is expected that this will be paid at the rate of £300 m. a year over the next five years.

The subsidization of passenger services may reduce road congestion and can be justified insofar as it corrects any distortion in the allocation of traffic caused by 'prices' for road passenger transport being below marginal social costs. The desire to avoid adding to unemployment by reducing the number working in British Rail may also be a factor of importance. But it would seem that the environmental benefits must form a major part of the supposed value received for the £392 million subsidy, however exaggerated a claim this may seem. Two important questions are at present unanswered. The first is whether the value of the social benefits received approximates to the subsidy costs incurred by the state. Since we have such a limited knowledge of the value of environmental benefits the question is likely to remain unanswered. Members of environmental groups and of pro-rail organizations like Transport 2000 are likely to put a high value on environmental benefits and to believe that the subsidies are justified. Others may be more doubtful. Frances Cairncross pointed out in a recent article in the *Guardian* that the expected rail subsidy needed for 1975 (£430 million) could also be used to build 20 new hospitals, and there are many alternative benefits which people might prefer to have instead of the relatively small environmental advantages provided by the railways. The payment of subsidies also has welfare implications which most economists would believe to be unfavourable since money raised from general taxation is used to help rail travellers who form a relatively 'better-off' section of the community.

The second question is whether some alternative use of the rail subsidy might not bring greater environmental and other social benefits (or equal benefits for a lower cost). This is a less difficult question to answer. As has been shown in other parts of this study there are numerous ways in which the environmental costs of road transport can be reduced. Since the roads carry by far the largest share of both goods and passenger traffic then there must be a strong prima facie case for believing that funds spent directly on reducing the environmental costs caused by this traffic will yield greater benefits to the community than will subsidies designed to keep rail traffic flows at an artificially high level. If rail subsidies were very small in relation to road expenditure then it would be possible to believe that the benefits from keeping traffic on the railways might be equal to the costs. But at the present level of rail subsidies (which have grown at a time when road expenditure has been reduced) it is extremely probable that greater environmental benefits would result from using these funds for such activities as providing roads which would take traffic away from people; designing and landscaping new roads in a way which would minimize their adverse impact on the environment; and finding methods of improving the technology of cars, lorries and aircraft so that they produced less noise and air pollution.

The extreme optimism of the proposals on rail subsidization contained in *Changing Directions* demands some comment. The 'Independent Commission' suggested that subsidies should be paid of 0·5p for each passenger mile above 18,000 million and 0·5p for each ton mile above 15,000 million.[1] If this advice had been followed in 1974 British Rail would have received total subsidies of only £6 million. The actual subsidy received would have required payment at a rate of 32·7p per passenger mile if the 'Independent Commission's' formula was used. The Commission's proposal was based on the belief that the subsidy would persuade the railways to obtain more traffic and, presumably, that average costs would fall dramatically with increasing traffic. But there is no evidence to support the curious belief that the railways could have won more traffic if they had only tried a little harder. The railways may, quite rightly, have rejected freight traffic when it was clear that revenue would fall short of marginal costs (and which would still provide inadequate earnings with a subsidy of 0·5p per ton mile), but there is no reason to suppose that the railway management was as spectacularly bad at salesmanship as the Commission seemed to imagine. Research carried out amongst firms in the west Midlands suggested that railway sales representatives were quite efficient and active. If the railways have losts good traffic to road hauliers and passengers to private cars it is for much more fundamental reasons than managerial inefficiencies. It is of course true that short-run marginal costs on rail may be very low, and that improvements in load factors would improve the financial position of the railways. But 'surplus' rail capacity is quite limited and may not be related to actual traffic flows. Substantial transfers of goods and passenger traffic to rail would improve but not transform rail finances and the 'Independent Commission's' subsidy proposals appear to be quite unrealistic. Suppose that, somehow, both rail passenger traffic and rail freight traffic carrying became twice as great as the figures shown in the 1974 accounts. Then the 'Independent Commission's' subsidy would amount to £159·8 million. But it involves taking an extremely optimistic view of the nature of railway costs to believe that doubling output would cut the amount of subsidy required to less than one half of its previous level. Significant increases in traffic would increase revenue, but they would increase costs as well. There does not seem to be any escape from the situation in which the price that must be paid for any environmental or other advantages that the railways can offer will be very substantial.

## Summary

*1. Road and rail are alternative techniques for providing transport services and, if possible, their comparative environmental advantages and disadvantages should be analysed without making any prior assumptions about the desirability of expanding (or contracting) the use of either mode of transport.*

*2. Except for a few special types of traffic, such as siding-to-siding bulk mineral traffic, railway technology is inferior to road technology as a means of meeting the present freight transport needs of Britain. Proposals to transfer goods traffic from road to rail must therefore be based on environmental and social considerations rather than on narrow 'economic' factors.*

*3. The transfer of goods traffic to rail can do little to reduce the environmental costs of road transport in urban areas where collection and delivery must be made by road.*

*4. Some case studies of possible road to rail transfers of general goods traffic showed that little environmental benefit would result. In some cases environmental costs might be increased by a traffic transfer.*

*5. Rail, with only 11,289 route miles of track open in Britain in 1974, does not have the physical capacity to carry more than a small proportion of the goods traffic now carried on the 212,034 miles of public roads open in Britain. In 1912, when over 20,000 miles of rail route were open, only 513·6 million tons of traffic were carried by rail while in 1974 the combined road/rail total was 1,881 million tons.*

*6. Railway operation itself produces such environmental costs as noise, visual intrusion and the severance of communities. Its main advantage is where a small number of trains can replace a larger number of lorries.*

*7. There would be environmental advantages in transferring passengers from a large number of cars into a few trains. Passenger traffic can be transferred to rail more easily than goods traffic because the 'transhipment' problem is much less serious. There may not, however, be any environmental benefits to be gained from transferring passengers from buses to trains.*

*8. It is uncertain whether rail is significantly more economical than road transport in using fuel in actual operating conditions. Its greatest advantage would be for bulk flows of mineral traffic which are already largely carried by rail. The price mechanism can be used to reflect the increased scarcity of petroleum-based fuels.*

*9. British Rail received government financial support of about £392 millions in 1974. Some of this subsidy may provide non-environmental benefits, but it is very doubtful whether the environmental advantages received even from the existing level of use of the railways could not be obtained more cheaply by alternative forms of public expenditure.*

Chapter ten

# Problems of town centres

## 1 Introduction

In this Chapter some of the environmental problems of town centres are discussed, particularly those relating to shopping. The Road Traffic Regulation Act, 1973, and the Town and Country Planning Acts of 1968 and 1971, gave local authorities wide powers to restrain traffic in order to improve the environment of town centres. The fundamental problem of town-centre shopping and business areas is the conflict between pedestrians and motor vehicles. In towns large enough to have separate central and suburban shopping areas it is likely that the majority of town-centre shoppers will need to travel by car, train or bus. A reconstruction of the 1972–3 National Travel Survey figures (in which percentage figures were related to mode rather than journey purpose) gives the approximate 'mode of travel' breakdown for 'shopping and personal business' for all households in the sample, shown in table 50. The percentages relate to passenger mileage.

Town-centre streets must also be used by delivery vehicles bringing goods to the shops, and, in many traditional 'High Street' shopping areas there is still a considerable volume of through traffic. This concentration of traffic has made many town-centre and urban shopping areas into unpleasant and dangerous places for shoppers and other users.

An obvious means of improving the town-centre environment is to develop traffic-free precincts where shopping crowds will not need to

**Table 50 Shopping and personal business travel, approximate percentages of passenger mileage travelling by different modes, National Travel Survey sample, all households, 1972–3**

| Mode of transport | Percentage |
|---|---|
| British Rail | 4·8 |
| London Underground | 0·5 |
| Local bus | 18·5 |
| Long-distance bus and coach | 0·6 |
| Works bus | 0·1 |
| Taxi | 0·3 |
| Car or van driver | 36·4 |
| Car or van passenger | 32·3 |
| Motorcycle | 0·5 |
| Cycle | 1·0 |
| Walk (over 1 mile) | 4·4 |
| Other private transport | 0·7 |
| Total | 100·1 |

Source: *National Travel Survey, 1972/3* (Department of the Environment, HMSO, 1975).

share road space with lorries, buses and cars. But equally obviously, if traffic is to be banned from a shopping street, then something must be done to find an alternative route for through traffic, to provide means for shoppers to reach the area, and to enable goods to be delivered to and from the shops. This means that if a precinct is to be completely pedestrianized, a by-pass or ring road must be built for through traffic, provision made for buses and car parks, and back access roads (or underground routes) provided for delivery vehicles. Many towns have now developed various types of shopping precinct. In this study we shall be mainly concerned with the problem of adapting traditional High Streets. Where a new town is built, or where there is a very comprehensive redevelopment scheme, it is now common practice to construct pedestrian-only precincts. This is a welcome development, but the problems involved relate mostly to town planning and will not be analysed here. Table 51 shows the traffic restraint situation in some towns in Britain.

## 2 The Watford precinct

Most of the discussion of the 'High Street' problem in this study is based on a survey which we carried out (together with David Whibley, now of the Polytechnic of Central London) for the Freight Transport Association, in 1972, of the Watford 'service-only' precinct.[3] Watford had a traditional High Street shopping centre, which carried both local

**Table 51 Traffic restraint in some town centres**

| | | |
|---|---|---|
| **1. Pedestrian precincts or foot streets at all times** | | |
| (a) *Associated with comprehensive redevelopment or similar schemes* | | |
| Scunthorpe | Denton, Lancs. | Swindon (part) |
| King's Lynn (part) | Bracknell (New Town) | |
| Lichfield | Harlow (New Town) | |
| Redhill | Stevenage (New Town) | |
| Stockport | Guildford (Friary St) | |
| (b) *Not associated with a redevelopment scheme* | | |
| (i) '*Wide*' *street situations* | | |
| Worthing | Southend | Eastbourne |
| Southampton | Hereford (2 schemes) | Hatfield |
| Rotherham | Swindon (part) | Bradford |
| Teesside–Middlesbrough | Bolton | Sheffield |
| (ii) '*Narrow*' *street schemes* | | |
| Bromsgrove | King's Lynn (part) | Havant |
| Norwich (London St) | Wallingford | |
| Horsham | Brixham | |
| **2. Pedestrian precincts or foot streets at certain times** | | |
| Maidstone (Week St) | Norwich (Red Lion St) | Guildford (High St) |
| Derby (Sadlersgate) | Swindon (part) | |
| Weymouth | Windsor | |
| **3. Limited access streets or area** | | |
| Leeds (at certain times) | Leicester | |
| Newcastle | Maidenhead | |
| Gateshead | Maidstone (Gabriel's Hill) | |
| Reading | Derby (parts of city centre) | |
| Hereford (1 scheme) | London (Oxford St) | |
| Birmingham | Watford | |

Source: E. Dalby, *Pedestrians and Shopping Centre Layout; a Review of the Current Situation* (TRRL Report LR 577, 1973).

and through traffic flows. The population of Watford was 78,117 in 1971, and that of the estimated shopping 'catchment area' was about 360,000. Over a period of about ten years Watford Borough Council developed the High Street, and some immediately adjoining roads, as a pedestrian-only area, although access is still allowed to buses and to delivery vehicles. The scheme has involved the construction of what is, in effect, an inner ring road which removes all through traffic, and of a number of car parks situated very near to the shopping area. Where possible back access roads have also been developed.

An attempt was made in our survey to summarize the gains and losses resulting from the scheme in a non-monetary form. (Our survey was made before the completion of the scheme. The main effect of its completion has been to make the environmental benefits more significant and to remove some of the disadvantages for through traffic.) The 'gains and losses' balance sheet was as follows:

*a.* *Delivery vehicles. Gain* in convenience for loading and unloading. *Loss* in causing longer journeys. Probable net *Loss.*
*b.* *Car shoppers' travelling time and convenience.* No significant difference.
*c.* *Rail and bus shoppers' travelling time and convenience.* Small *Loss.*
*d.* *Non-shoppers' travelling time and convenience.* Small *Loss.*
*e.* *Shopkeepers' delivery vans.* Small *Loss.*
*f.* *Walking distances.* No significant difference for car shoppers. Small *Gain* for lorry drivers.
*g.* *Through traffic.* No significant difference.
*h.* *Environmental factors.* Large *Gain.*
*i.* *Accident rates. Gain.*

This analysis avoids the basic problem of evaluating gains and losses and of calculating an overall balance of advantage. Consideration of all the factors in the Watford example suggest that the scheme has produced considerable benefits for the community. This does not necessarily imply that it was justified as greater benefits might have been obtained from some alternative form of investment. The loss caused to delivery vehicles by increasing the length of their delivery journeys will have become more significant with rises in fuel prices.

## 3 Requirements for environmental improvement of town centres

Key factors in the success of the Watford scheme, and of some similar developments, have been the ability to divert through traffic on to some kind of ring road system, and to provide car parking space very near to the shopping centre. Towns which cannot construct car parking facilities near to their central shopping area are unlikely to prove attractive for general family shopping in the future. Food purchases and general household goods are bulky and heavy and it is difficult to see how even the most efficient system of public transport can compete with the private car in carrying these types of goods. Private cars may be inefficient in using road space when they carry one person to work. But for family shopping, where both people and goods are carried, and where the goods must be delivered to houses, they must often be the most efficient form of transport available. There are many towns in Britain which differ from Watford in some essential respects and where the solution of the problem of the town centre may be different.

Very large towns which are regional shopping centres will be able to attract large numbers of people for 'specialist' shopping where they can enjoy a wide range of choice. These shoppers will mostly be prepared to use public transport or to drive to car parks which are not immediately adjacent to the main shopping streets. As has been shown in Leeds, regional shopping centres can close some shopping streets to cars with little danger of discouraging shoppers. The main problems in the largest towns are in providing alternative routes for traffic displaced from the 'pedestrianized' roads. This problem is particularly acute in London where the main shopping streets often carry heavy traffic streams which cannot easily be diverted. Several towns, like Leicester, allow buses to continue to use roads which have been closed to cars. This can seriously reduce the environmental benefits of pedestrianization schemes, and increases accident risks.

The town-centre problems may be most difficult to solve in smaller towns which may not be able to maintain a viable shopping centre for specialist non-grocery shopping only. This kind of shopping demands a minimum size of catchment area. They may also find it difficult to attract branches of some of the large national stores. It was found during the Watford survey that one shop belonged to a company whose national policy was not to open any branches with less than 15,000 sq.ft. (1,393·5 sq. metres) of sales room. These smaller towns must thus continue to rely on all-purpose shopping. If they adopt pedestrianization schemes which will make it difficult for people to bring their cars within very easy reach of the shopping streets then they will probably lose shoppers, who will choose centres where they can do all-purpose shopping which they can load into the family car. Towns which do not have a catchment area large enough to sustain a specialist shopping centre and which are prevented by their geography from clearing areas near the central shopping streets to provide for through traffic and to provide car parking space may therefore be unwise to develop a too ambitious pedestrianization scheme. This is of course a generalization. Isolated towns where there are no rival shopping centres available for shoppers will obviously be in less danger of losing trade than those in areas with numerous shopping centres within easy reach by car. The problem may be particularly great in towns which have many beautiful or historic buildings and which cannot be 'developed' with car parks or ring roads. These towns may, however, be able to attract specialist and tourist trade, and to develop other attractions such as good restaurants which will keep their central areas 'alive', even if some shopping trade is lost.

One problem associated with all town-centre development and improvement schemes is that of the level of expenditure. In some cases a 'second-best' scheme has to be adopted, not because it would be physically impossible to provide a better one, but because of cost constraints. An ideal scheme would provide for a complete separation between people and vehicles. Buses would not need to use the pedestrianized shopping streets and delivery vehicles would be provided with separate access to shops, either by back access roads or by underground tunnels. The 'fair shares for all principle' or 'spatial equality' as it is called in some local authority TPPs, may mean that no town can develop a really satisfactory scheme. While there are obvious political difficulties in telling the people of Little Backwater that they must leave their 'High Street' unimproved while their neighbours in Great Backwater carry out a major pedestrianization scheme, the community might nevertheless benefit from having one good scheme rather than two imperfect ones.

## 4 Goods vehicles and town-centre shopping

The problem of the 'visual' intrusion of goods vehicles has already been discussed in Chapter Seven. This may be particularly important in town-centre shopping areas. Only relatively few schemes (usually those in New Towns and other totally new developments) of pedestrianization have succeeded in removing delivery vehicles completely from the shopping area. Even in these cases goods vehicles may have to go through unsuitable roads to reach the back access routes to the shops. Towns with shopping streets still open to traffic will of course find that goods vehicles are particularly intrusive, especially as they have to park in order to make deliveries. Two possible methods of dealing with this problem have been developed or considered. Deliveries to town-centre shops can be greatly reduced if only one sample of each type of good is stocked, and deliveries to customers are made from warehouses situated outside the town-centre. This is, of course, only feasible for large goods which customers do not wish to carry away themselves. Some large stores in the Watford survey had warehouses of this nature. Although this system may reduce the congestion in central urban areas it has some disadvantages. The total expenditure of resources on transportation may be increased because delivery to customers' homes is implied. Shoppers taking goods home themselves use their cars (or public transport) as delivery vehicles at a very low, or zero, marginal cost. Delivery from warehouses is also only possible with durable goods of a type which shoppers would not prefer to carry away themselves.

Another method for reducing the intrusion of goods vehicles into town-centres is to use transfer depots. These could have two advantages. First, goods could be transhipped from large vehicles to smaller lorries or vans which would be less intrusive in town-centre areas. Second, if deliveries for a number of shops were carried in one town-centre delivery vehicle, the total numbers of vehicles running in the central areas might be considerably reduced. Interesting examples of this approach are the large privately organized transfer depots established to the north and the south of Paris. Unfortunately their success has been limited, with costs higher than anticipated. The largest transfer depot has been used more as a warehouse than as a transfer point for breaking down the loads of large lorries into smaller loads for delivery by small vans within Paris. Other schemes have been more successful. In Holland there are 27 transfer depots owned jointly by road hauliers, and these have both cut costs and reduced the environmental disbenefits caused by heavy vehicles. In Britain, RHM (the millers and bakers) following the introduction of transhipment, reduced its number of delivery vehicles from 363 to 266, and halved the number of calls made by each vehicle. Larger companies are able to support investment in a wider network of warehouses. In the U.K. a typical eight-acre depot costs approximately £2 million and building costs may be £10 per square foot.[3] The distribution company of Unilevers', SPD, has 30 warehouses and is therefore able to operate with smaller vans over shorter distances. By contrast smaller hauliers without warehousing facilities use larger vehicles over longer trunk hauls. The shopping centre environment could be improved by a network of government-sponsored warehouse facilities, although as British Rail has shown, providing subsidies without being able to quantify the benefits is a dangerously 'soft' option.

One of the most comprehensive studies of an interchange depot is that recently carried out for Chichester in Sussex.[5,6] The financial, operational, and environmental implications of a range of options were considered, depending upon whether or not the depot was to handle all goods or a limited range of goods, the carrying capacity of a special depot fleet, and frequency of delivery. The consultants engaged to examine the financial and operational implications of a depot concluded that only a limited number of the retail goods delivered into the central area would be suitable for transhipment through a local interchange depot, representing approximately 50 per cent of the total throughput of retail goods to the central area. Secondly they argued on organizational efficiency grounds that a depot-based fleet of vehicles is preferable to traders collecting their own goods from the depot. They examined two possible carrying capacities for a special depot fleet, 1·5 tons and 3 tons. All the options using the smaller vehicles were dismissed on grounds of excessive cost. Finally, a high level of service would benefit manufacturing, distribution service companies and traders, given lower stock levels and lower lead times between order booking and delivery. The environmental implications of the various options were assessed by the County Planning Officer. His report favoured the depot handling all goods since this would make it possible to remove all existing heavy delivery vehicles. The environmental assessment recommended a special depot fleet of 1·5 tons capacity vehicles on the grounds that they are smaller and less obtrusive in the urban setting of the small side streets of the city, despite the larger number of vehicular movements involved.

Whatever 'option' was selected would involve a financial cost, occasioned by the creation and operation of the depot, and the double-handling of goods necessarily involved. The trade-off with the environmental evaluation is illustrated by constructing an index to show the cost-effectiveness of reduced volumes of servicing traffic. Table 52 shows the annual cost (row 1) of the depot options recommended by the consultants (columns 3 and 4), and those which offer the greatest environmental benefit (columns 1 and 2). This is divided by the average number of vehicle movements per annum which would be saved (rows 2 a–d), and the resulting index for each group is set out in rows 3

**Table 52 Cost effectiveness of preferred depot options**

| Option | 1 (L.3d full) | 2 (L1.5d full) | 3 (H.3d Ltd) | 4 (L.3d Ltd) |
|---|---|---|---|---|
| 1. Annual financial cost (£) | 302,480 | 357,135 | 120,805 | 116,405 |
| 2. *Reduction* in number of vehicle movements per annum by: | | | | |
| (a) Vehicles of all weights | 271,320 | *127,120* | 143,360 | 165,760 |
| (b) Vehicles above 1·5 tons | *84,280* | 226,800 | *36,120* | *57,400* |
| (c) Vehicles above 3 tons | 193,200 | 193,320 | 117,320 | 117,320 |
| (d) Vehicles above 5 tons | 25,200 | 25,040 | 12,040 | 12,040 |
| 3. Cost effectiveness of reducing: | | | | |
| (a) Vehicles of all weights | 1·11 | 2·18 | 0·84 | 0·70 |
| (b) Vehicles above 1·5 tons | 3·59 | 1·57 | 3·34 | 2·02 |
| (c) Vehicles above 3 tons | 1·57 | 1·84 | 1·03 | 0·99 |
| (d) Vehicles above 5 tons | 12·00 | 14·26 | 10·03 | 9·67 |

*Notes*

1. L.3d full = full throughput, low frequency of deliveries, 3-ton load vehicles
   L1.5d full = full throughput, low frequency of delivery, 1·5-ton load vehicles
   H.3d Ltd = limited throughput, high frequency of deliveries, 3-ton load vehicles
   L.3d Ltd = limited throughput, low frequency of deliveries, 3-ton load vehicles
2. County's estimate of vehicle movements/day factored by 280 to give annual vehicle movements.
3. The depot vehicle in columns 1, 3 and 4 is introduced at 2(b); in column 2 it is introduced at 2(a).

Source: *Chichester Central Area Servicing System*, Local Interchange Depot Study (Nathaniel Lichfield and Partners, 1975).

a–d. The full depot options give lower cost effectiveness than do the limited throughput options, but the latter have a smaller environmental advantage. The consultants' recommendation of 3-ton capacity vehicles to deliver a limited range of goods would require a fleet of 13–14 vehicles, and the employment of about 38 staff on a full-time basis. The depot would occupy a site of not more than 1·7 acres.

Many problems arise from the study. For example, would an out-of-town shopping centre be a more socially advantageous use of 1·7 acres.[7] How will existing through traffic be affected? Would the results be sensitive to an evaluation of a wider range of carrying capacity for a special depot fleet? Would the problem of non-compliance be significant? A study carried out in Oxford of the observance of a ban prohibiting vehicles over 3 tons unladen weight from entering the city unless they required access, found that only about 2 per cent of all heavy goods vehicles entering the restricted area contravened the ban.[7] A similar study at Newbury, however, found a higher proportion guilty of non-compliance.[8] The consultants suggest that the central area could be placed under a 'Dykes Act' restriction, although they concede that this would have limited effect because the lowest maximum weight restriction that can be imposed would still permit many existing deliveries to be made direct to outlets. Also given the wide variety of vehicle shapes and sizes for any given weight of vehicle, a weight restriction would not necessarily eliminate the bad effects of goods vehicles on the urban fabric of historic towns. Finally, the operation of the consultants' recommended options would imply an increase of about £4 in handling costs per ton of goods. The county's environmental study sets out, but does not quantify, the environmental implications of the reduced incidence of vehicles. Would consumers of any goods bearing an increase in price because of increased distribution costs be willing to pay an extra £4 per ton if asked? Should the general taxpayer, through Government subsidy, support the costs of improving the environment of people at Chichester? These problems still remain.

## 5 Out-of-town shopping

The final question to be considered is how far the preservation of town centres as major all-purpose shopping areas should be considered to be an overriding objective. There can be little doubt that many town centres should remain places which attract people, and that urban decay is very undesirable. But this does not necessarily mean that all town centres must cater for all kinds of shopping. As already argued, some central areas can flourish by providing specialized shopping, restaurants, and entertainment. If some British cities have managed to retain more vital central shopping areas than their American counterparts, they also suffer from partial desertion at night when all the shops are closed. There may even be some town centres which are not worth preserving as shopping centres at all and which could be left to become mainly business centres.

As an exercise in distribution, town centre shopping does not appear to be particularly sensible. Most 'High Street' shopping areas developed originally when people lived very near, and could walk to their High Street to shop and find their entertainment. But the suburban sprawl of modern cities has changed this situation. Goods are now brought into the town centres, through the suburbs, to be collected by people who also live in the suburbs, and who will carry them out to the suburbs again. Interchange depots of the type discussed in the Chichester study, would be built just outside town centres and would use up considerable space. It would obviously be more efficient to allow shoppers, who also live mostly outside the town-centre areas, to drive or walk into these depots and collect their goods themselves. In other words the depots could be replaced by new suburban shopping developments or by an extension of the central shopping area. In some areas out-of-town or edge-of-town shopping centres would provide the most efficient method of meeting people's shopping needs. Distributive efficiency is, of course, only one consideration, and, as we have argued in the rest of this Chapter, the preservation or development of town centres which are full of life, interest, and activity is an extremely important aim. The exercise of very careful control over the development of out-of-town or large suburban shopping areas is certainly justified. But where a town-centre shopping area can only be preserved by such an elaborate device as the interchange depot the alternative of suburban or fringe shopping centres should be considered. The artificial preservation of all-purpose town-centre shopping by a rigid use of planning controls is not always automatically justified. A more positive approach may be to find other activities such as specialist shopping, various forms of entertainment, and facilities for eating and drinking, which will keep town centres alive. Where town-centre shopping areas cannot be adapted to modern needs, attempts to resist change may destroy what was intended to be preserved. Where town-centre all-purpose shopping can only be continued at the cost of having streets crowded with all kinds of vehicles then maybe it should be abandoned.

## Summary

*1. Most town centres in Britain were not developed to meet the demands of motorized transport. The conflict between people and traffic greatly reduces the attractions of town centres as places for shopping and entertainment.*

*2. In order to be able to develop existing 'High Streets' into traffic-free areas which will attract all-purpose shoppers it is necessary to be able to provide alternative routes for diverted traffic and to provide ample car parking space and good public transport access close to the shopping area.*

*3. Towns with large catchment areas or with some special attractions may be able to develop flourishing specialist shopping centres. For this kind of shopping people will travel by public transport and nearby car-parking space is not essential. It is thus relatively easy for these towns to develop high-amenity, pedestrian-only central shopping areas.*

*4. There are some 'High Streets' which are not particularly attractive, which cannot easily be developed as modern all-purpose shopping centres, and which are unlikely to be viable as specialist shopping centres. In these cases there is no good reason for resisting the development of suburban or out-of-town shopping centres.*

*5. It seems unlikely that the construction of interchange depots for goods vehicles will be justified except in a few special cases. The transportation of large quantities of goods traffic and of people into inner-urban areas which are not equipped to handle them (except at great cost to the environment) and the subsequent transportation of these same goods and people back into suburban areas is an obviously wasteful procedure.*

Chapter eleven

# Conclusions

Each Chapter in this study has been concluded with a summary of the main points of the discussion. These final conclusions are therefore restricted to a discussion of ten of the more important issues which relate to government policy and to choices and decisions which must be made in the future. In some cases they represent our own opinions on these issues.

## 1 The problem of choice

All, or almost all, environmental improvements will involve costs for the community. If, as a nation, we want to enjoy an improved environment then we shall have to make some corresponding sacrifices of alternative forms of satisfaction. The costs of reducing the level of environmental pollution caused by transport operations cannot be made to be borne exclusively by road hauliers or airline operators or any other providers of transport services. The community, through its demand for transport, pollutes itself, and must itself ultimately pay the costs of improvement. If we want quieter or smaller lorries then we must pay more for our groceries, and if we want safer roads we must be prepared to take longer over our journeys or to allocate more funds to pay for the improvement of accident black sites. There has been a very strong demand in Britain, and in many other countries, in recent years to reduce the environmental costs of transport. The fundamental problem for government is to decide how great an improvement is required, and to try to distinguish the point at which the value of the benefits from further environmental improvements will be less than the cost of bringing them about. At the moment there is a great deal of ignorance about how much these improvements are worth, and a lesser degree of uncertainty about the costs. Measurement of the value of benefits is very difficult since most of these benefits cannot be sold on the market. There may be very great differences between individual valuations, and value judgments are necessary in order to decide which measures will yield net benefits for the community. There is considerable evidence to suggest that environmental improvement is a 'luxury' good and that richer people will be prepared to pay more for each unit of improvement than will those with lower incomes. Our lack of knowledge means that policy decisions such as a reduction in maximum permitted vehicle noise levels, or the alteration of speed limits, must be made, which may or may not yield a net benefit for the community.

## 2 The demand for transport

The pollution caused by transportation, like other forms of environmental pollution, is closely associated with the level of economic activity. An increase in the economic growth rate will be likely to lead to an increase in the environmental costs of transport (although it may also make it possible to raise the level of investment in measures designed to combat environmental pollution). The problem of how far economic growth should be deliberately restricted in order to preserve the environment is outside the scope of this study. But there are some ways in which the effective demand for transportation could be reduced without necessarily reducing the size of the gross national product. These measures also involve complicated considerations and could increase some (non-transport) environmental costs while reducing those caused by transportation.

There are two short-term and two long-term measures which would reduce the effective demand for goods and passenger transport services. Some small reduction in the demand for ton miles of goods transport could result from rationalizing the delivery of consumer goods, but this might involve a reduction in consumer choice. In the long run, new industrial location could be concentrated around rail and motorway access points, but this might restrict the freedom of people to live in areas of low density development. It would, to some extent, involve returning to high-density urban development of the mid-nineteenth-century pattern. The demand for passenger transport could be reduced if people chose to live nearer their place of work. A very crude examination of the 1966 Sample Census statistics suggests that some rationalization of the pattern of commuter journeys might be possible without any significant loss of amenity. This is particularly true in the large conurbations where considerable numbers of people appear to live in area 'X' and work in area 'Y', while equal numbers of people live in 'Y' but work in 'X'. It is easy to understand why people who have to work in Darlaston might choose to live in Sutton Coldfield, and it is obvious that most of the people who work in the City of London must of necessity live elsewhere. But it is not so clear why, for example, an almost equally large number of people should choose to live in Ealing and work outside as choose to live outside Ealing but work inside that borough. It may be that (at least until recently) public transport has been provided too cheaply in the large conurbations because the environmental implications of encouraging commuter traffic have not been taken into consideration.

In the long run there would be considerable reductions in the demand for costly passenger transport if people lived in smaller towns. Cities like London, Birmingham and Manchester are a 'mistake' from the transport point of view. People are forced to use motor transport, which creates environmental costs and consumes fuel, unnecessarily. People may, of course, gain some satisfaction from living in very large cities which outweighs the transport environmental costs. But the replacement of the conurbations by relatively small towns would undoubtedly do a great deal to reduce the environmental costs of transportation.

The effective demand for goods transport could also be reduced by using larger vehicles, although some people believe that the extra environmental costs of the heavier vehicles would outweigh the benefits. Our conclusion is that there would be net environmental benefits resulting from allowing some increase in vehicle weights. This issue is, however, partly dependent on what provision is made for new road building. It is a commonplace that the use of public transport can reduce the effective demand for transport. A bus with 45 passengers will almost certainly be less costly to the environment than 35 or 40 cars carrying the same passenger flow. Similarly a commuter train full of passengers will usually do less damage to the environment than the same number of people travelling in cars with an average load of not much more than one person per car. For goods traffic the same principle applies, though it is less significant. The most important form of public goods transport is that provided by road hauliers (and not rail goods transport, as is sometimes supposed). The effective demand for transport will usually be reduced if goods are amalgamated into full lorry loads, on both outward and return journeys, by road hauliers. Own-account operation will often (though not always) involve running a greater vehicle mileage. But there may sometimes be important advantages in using private cars or own-account goods transport which offset their environmental disadvantages.

### 3 Use of the price mechanism

Where it is possible to adopt the 'polluter must pay' principle and compel the agency causing transport environmental costs to pay to reduce them, this can lead to an efficient solution to the problem. This remains true whether the immediate 'polluter' pays the cost himself or, as is more likely with transport environmental costs, passes them on to ultimate consumers. But the use of the price mechanism to 'charge' for transport environmental costs is seldom possible. As has already been argued we usually only have a very imperfect knowledge of the money value of environmental costs. A tax on the creation of environmental disbenefits would also often be extremely complicated if it was to be at all closely related to the actual creation of these costs. Two lorries might produce exactly the same noise level, but if one operated almost exclusively on motorways while the other ran mostly in built-up areas the costs which they imposed on the community would be very different. But there is one area in which the price mechanism could be used more effectively without any great difficulty. This is in charging for road wear and tear. As was shown in Chapter Five, the present vehicle licence duty which varies with unladen weight is a particularly inefficient way of charging lorries for the road wear which they cause. A road wear and tear tax was developed which reflects gross weight, the number of axles and average annual vehicle mileage.

### 4 Accident evaluation

The money value put on the cost of injury or fatal accidents is a highly emotive and difficult subject, and we would not claim to have produced any figures which are better than those issued by the Department of the Environment. It does seem, however, that this is an area in which more public involvement might be possible. The figures fed into the 'COBA' analysis of road investment proposals or used in evaluating specific proposals which will reduce road accidents involve philosophic assumptions which are certainly not generally understood by all elected representatives or officials or by the general public. It is difficult to challenge figures which are fed into a complicated computer programme. It may well be that people would be prepared to spend more on reducing accidents than their observed behaviour seems to suggest. The method of dealing with accident 'black sites' seems to be particularly unsatisfactory.

It should be possible to justify expenditure on potential 'black sites' and not to confine investment to those which have earned their status because a number of accidents have already occurred.

## 5 The road and rail solutions

There is little doubt that the most unpopular conclusions of this study will be those which relate to the use of road and rail transport. The evidence seemed to us to show quite clearly that there would be few, if any, environmental gains resulting in transferring goods traffic from road to rail. The arguments set out in Chapter Nine need not be repeated. The most conclusive is that large-scale transfers would be physically impossible unless there was to be massive investment in the construction of new rail track. There will be environmental advantages in transferring some passenger flows from car to rail, although even here rail capacity is limited and transfers from cars to bus transport may be at least equally advantageous. New road construction, on the other hand, seems to us to be the most hopeful method of reducing transport environmental costs in the next few years. The great advantage of road construction is that total traffic flows can be removed from areas where they harm the environment to those where these costs will be minimized. Many sincere environmentalists seem unable to believe that any conclusions which favour road rather than rail transport can be based on independent research. A recent publication by the Conservation Society called *Roads to Ruin* considered that people who argued for any road building must have an ulterior motive. 'Some of these people make money out of roads and road transport – all see their careers, power and prestige enhanced by massive road construction'. Our only 'vested interest', like most people in Britain, is that of taxpayers, transport consumers and people who live in the environment of modern Britain. People like the 'M69 Now' group who campaigned against the postponement of the building of the M69 motorway because they were anxious for traffic streams to be removed from the towns and villages where they lived, had an appreciation of the realities of the situation. New roads, particularly by-passes, are needed to meet existing traffic needs and will be required even if official estimates of the growth in the demand for goods and passenger traffic prove to have been exaggerated. If, as is sometimes claimed, goods traffic were generated by the provision of new roads then an easy solution to the problem of Britain's low growth rate would be available. The demand for goods transport can only increase when gross national product increases and there are more goods to be transported. Roads do have environmental costs of their own though these can be minimized by proper landscaping and planning, and there are, of course, many places in urban areas where the objective of separating traffic from people cannot be achieved.

It is not the purpose of this study to discuss the future of the railways in Britain, but the arguments of the book do suggest that large subsidies to the railways may not be justified on environmental grounds. It should be made clear, however, that the analysis in this study relates to the proposals to transfer traffic from road to rail. We have not attempted to measure the environmental impact of reducing the scale of rail operations and transferring goods and passengers from rail to road transport. In the long run, when supplies of petroleum-based fuel are exhausted, and if satisfactory electric-battery-operated vehicles have not been developed, it may be necessary to provide electric power directly to road vehicles from overhead cables. The reintroduction of trolley buses and the development of 'trolley lorries' might become necessary. Even if all the existing rail track were electrified, British Rail would be quite incapable of meeting the transport needs of the twenty-first century, and building new electric rail lines would probably be more costly than providing electric power for road traffic.

## 6 Compensation

Investment in transport infrastructure such as roads, docks and airports is often delayed by the violent objections of those who will suffer from the new developments. The principles of the Land Compensation Act could be greatly extended in future and generous compensation paid to all those who will suffer loss of amenity because of the completion of new transport works. In many cases this would both be equitable and lead to increased economic efficiency (because the costs of compensation would be less than the value of the gains to the community from reducing delays in the investment programme).

## 7 The European Economic Community

Britain's membership of the EEC has in some ways made it much more difficult to resolve many of the policy problems relating to transport and the environment. It may be desirable to harmonize environmental standards in all EEC countries, but the process of reaching agreement on the issues involved is likely to prove extremely difficult. If it is hard to resolve environmental problems which involve complicated economic and political considerations in one country, the difficulties are likely to be magnified where nine governments must agree. Current issues which have to be resolved, and which have environmental implications include those of the maximum weights of goods vehicles; the use of tachographs; the limits on driving hours for public service vehicles and goods vehicles; and regulations relating to vehicle licensing. We believe that Britain should accept higher vehicle weights and the use of traffic tachographs, but that the policy of reducing drivers' hours needs to be justified by an examination of the effect on accident rates. A reintroduction of the legal distinctions between 'own-account' and 'hire or reward' operation can only have harmful environmental and economic effects. It would be absurd to increase total vehicle mileage and fuel consumption by preventing own-account operators from making an optimum use of their vehicles. Current EEC tax proposals for goods vehicles would, if adopted, make lorries with a larger number of axles relatively more attractive to operators.

## 8 Town-centre shopping

Although we recognize the importance of preserving the vitality of town-centre areas, this does not mean that all out-of-town or suburban shopping development is undesirable. Some existing 'High Streets' cannot be made suitable for all-purpose shopping and are unlikely to be viable as centres for specialist shopping.

## 9 Canals and other forms of transport

There are those who believe that a major contribution towards solving the problems of transport and the environment could be made by revitalizing the canal system, or by the use of some largely untried form of transport such as airships. Nothing has been said about these proposals in this study because we were not aware of any adequate research data which could be used to evaluate them. This does not mean that new, or revived, technologies should be ignored. As long ago as 1895 plans were drawn up to build canals from London to Birmingham and from Birmingham to Liverpool which would be able to carry bulk traffic flows. The Birmingham/Liverpool canal would, for example, have carried sea-going vessels of 300 tons, or 400-ton barges. There is a good case for carrying out research into the costs and benefits of constructing canals or investing in new forms of transport, and trying to measure the environmental advantages and disadvantages of such proposals.

## 10 Public opinion

Most decisions relating to measures intended to reduce the environmental impact of transport operations involve highly complex analyses and to some extent may depend upon 'guesses' about what will happen in the future. It would be unrealistic to suppose that the 'man in the street' can be involved in the details of these decisions. But we believe that it is important that the basic assumptions underlying investment and other decisions should be made clear and that there should be both opportunities for the expression of public opinion and regular attempts to find out what people are thinking. We do not believe that economists will be able to produce an 'optimum' solution to environmental problems merely by studying the ways in which people choose to spend their incomes, nor do we believe that technological developments alone will provide satisfactory solutions. In the expression of public opinion we believe that the environmental 'pressure groups' have performed a valuable function in drawing attention to environmental problems. But we consider that the emphasis should now be shifted from protest to co-operation in attempts to find some answers to the problem of the environmental costs which result from transport operations.

# Additional bibliography

Most of the books and articles used in making this study have been included in the list of references. But there are a few general works to which no specific reference has been made but which readers who wish to discover more about the environmental problems of transport may find to be of value. These are listed below (some works which were not available until after this study was completed are also included).


B. T. Bower and A. Kneese, *Environmental Quality Analysis* (John Hopkins, 1972).

P. F. Bowers, *Private Choice and Public Welfare* (Dryden Press, 1974).

M. Edel, *Economics and the Environment* (Prentice-Hall, 1973). (See particularly Chapter six on 'The Social Organization of Internal Combustion'.)

A. V. Kneese, S. E. Rolfe and J. W. Harned (eds.), *Managing the Environment* (Praeger, 1973).

A. Lassiere and P. Bowers, *Studies on the Social Costs of Urban Road Transport*, European Conference of Ministers of Transport 18th Round Table, 1972.

C. A. Nash, *Public versus Private Transport* (Macmillan, 1976).

D. W. Pearce, *Enviromental Economics* (Longman, 1976).

R. Pryke and J. Dodgson, *The Rail Problem* (M. Robertson, 1975).

J. G. Rotherberg and I. G. Heggie (eds.), *Transport and the Urban Environment* (Macmillan, 1974). (See particularly Chapter 6, 'Transport and the Urban Environment', by C. D. Foster.)

C. K. Rowley and A. J. Peacock, *Welfare Economics, A Liberal Restatement* (M. Robertson, 1975). (Contains a critical analysis of Paretian welfare economics and an attempt to work out 'Liberal' alternatives. Chapter 8 on 'Market Regulation' is of special interest.)

# Additional bibliography

Transport Policy, A Consultation Document, Vol. 1 and Vol. 2 (HMSO, 1976). (Contains a statement of government policy on transport problems and invites discussion.)

I. Walter, *International Economics of Pollution* (Macmillan, 1975).

R. O. Zerbe and K. Croke, *Urban Transportation for the Environment* (Ballinger, 1975). (Despite the title this book deals exclusively with air pollution in the U.S.)

# References

## Chapter one


1. A. A. Walters, *Noise and Prices* (1975).
2. *Noise in the Next Ten Years* (The Noise Advisory Council, HMSO, 1974).
3. A. K. Dasgupta and D. W. Pearce, *Cost-Benefit Analysis: Theory and Practice* (1972).
4. W. J. Baumol and W. E. Oates, *The Theory of Environmental Policy* (1975).
5. E. J. Mishan, 'Pangloss on Pollution', *Swedish Journal of Economics, 73* (1971).
6. R. Millward, *Public Expenditure Economics* (1971).
7. D. M. Winch, *Analytical Welfare Economics* (Penguin edn, 1971).
8. J. E. Meade, *The Theory of Economic Externalities* (Holland, 1973).
9. R. Turvey, 'On Divergencies between Private and Social Cost', *Economica*, August 1963.
10. *The Polluter Pays Principle* (Organisation for Economic Co-operation and Development, 1975).


## Chapter two


1. A. H. Tulpule, *Trends in transport of freight in Great Britain* (Road Research Laboratory (hereafter RRL) Report LR 429, 1972).
2. J. C. Tanner, *Forecasts of vehicles and traffic in Great Britain: 1974 Revision* (TRRC Labs. Report 650, 1974).
3. *National Income and Expenditure, 1972* table 15.
4. *Censuses of Production*, 1958, 1963, 1968 and 1970.
5. *Survey of the Transport of Goods by Road, 1967–68* (Department of the Environment, 1971).
6. *The Transport of Goods by Road: 1970–72* (Department of the Environment, 1972).
7. *Chichester Central Area Servicing Systems*, Local Interchange Depot Study, Final Report (Nathaniel Lichfield and Partners, 1975).
8. Clifford Sharp, *The Allocation of Freight Traffic – A Survey* (Report to Ministry of Transport, 1970).
9. A. Brown and A. Maultby, 'Methods of forecasting the quantities of freight in Great Britain', *Statistical News, 25* (May 1974).

10. A. Tulpule, *An analysis of some world transport statistics* (Transport and Road Research Laboratory (hereafter TRRL) Report LR 622, 1974).
11. *National Travel Survey, 1972/73* (Department of the Environment, HMSO, 1975).
12. *Passenger Transport in Britain* (Department of the Environment, formerly Ministry of Transport; HMSO, various years).
13. K. M. Gwilliam and P. J. Mackie, *Economics and Transport Policy* (1975), Chapter 4.
14. *Sample Census 1966*, Workplace and Transport Tables, Part II (HMSO, 1968), and 1971 *Census*, Workplace and Transport Tables, Part II.
15. Clifford Sharp, *Problems of Urban Passenger Transport* (1967).
16. R. Smeed and J. Wardrop, 'An Exploratory comparison of the advantages of cars and buses for travel in urban areas', *Institute of Transport Journal*, March 1964.

## Chapter three

1. *A review of road traffic noise* (RRL Report LR357, 1970).
2. R. Stephens and A. Bate, *Acoustics and Vibrational Physics* (1966).
3. J. Blitz, *Elements of Acoustics* (1964).
4. A. Wood, *Acoustics* (1950).
5. *Code of practice for reducing the exposure of employed persons to noise* (Department of the Environment, HMSO, 1972).
6. P. M. Nelson, *A computer model for determining the temporal distribution of noise from road traffic* (TRRL Report LR 611, 1973).
7. N. F. Ross, 'Noise from light vehicles', Seminar on the Propagation of Traffic Noise, 1971 (unpublished).
8. P. T. Lewis, 'The noise generated by sample vehicles in freely flowing traffic on level roads', Seminar on the Propagation of Traffic Noise, 1971 (unpublished).
9. A. W. Christie, J. Prudhoe and M. A. Cundill, *Urban freight distribution: a study of operations in High Street Putney* (TRRL Report LR 589, 1973).
10. A. W. Christie, R. S. Bartlett, M. A. Cundill and J. Prudhoe, *Urban freight distribution: studies of operations in shopping streets at Newbury and Camberley* (TRRL Report 603, 1973).
11. D. Starkie and D. Johnson, *The Economic Value of Peace and Quiet* (1975).
12. Professor T. Priede of the Institute of Sound and Vibration Research, lecture given at Loughborough University of Technology, 8 March 1972.
13. P. E. Waters, *Commercial Vehicle Exterior Noise* (Institution of Sound and Vibration Research).
14. *Traffic Noise: The Vehicle Regulations and their Enforcement* (Working Group of Noise Advisory Council, 1972).
15. D. T. Aspinell and J. West, *The Reduction of the External Noise of Commercial Vehicles by Engine Enclosure* (MIRA Report 17, 1966).
16. T. Priede, 'Diesel engine noise control in the 1970s', paper presented at the Automobile Engineers' Conference on Noise, Dirt and the Diesel, March 1972.
17. N. Lalor, paper delivered to British Association, September 1972.
18. P. M. Nelson and J. Fanstone, *Estimates of the reduction of traffic noise following the introduction of quieter vehicles* (TRRL Report 624, 1974).
19. J. Dunn, *Traffic Census results for 1972* (TRRL Report 618, 1974).
20. A. Alexandre and J. P. Barde, *Le Temps du Bruit* (Paris, 1973).
21. Commission on the Third London Airport, *Papers and Proceedings* (1970).
22. W. L. Baxter, 'Legislative influences on commercial vehicle design in Europe',

paper delivered at 1972 National Conference of the Institute of Road Transport Engineers.
23. *Which?* (Consumers' Association, August 1974).
24. C. Sharp, *Living with the Lorry* (1973).
25. Jay Diffey, 'An investigation into the effect of high traffic noise on house prices in a homogeneous sub-market', paper presented at Centre for Environmental Studies seminar on House Prices and the Micro-economics of Housing, December 1971.
26. Robin Toune, 'An investigation of the effect of freeway traffic noise on apartment rents', cited in n. 23 above.
27. M. E. Burt, *Roads and the environment* (TRRL Report LR 441, 1972 ).

## Chapter four

1. A. W. Christie, J. Prudhoe and M. A. Cundill, *Urban freight distribution: a study of operations in High Street Putney* (TRRL Report LR 586, 1973); A. W. Christie, R. S. Bartlett, M. A. Cundill and J. Prudhoe, *Urban freight distribution: studies of operations in shopping streets at Newbury and Camberley* (TRRL Report 603, 1973).
2. *Environmental Bulletin* (Department of Trade and Industry), no. 13 (October 1972).
3. *Technology and the Environment* (Department of Trade and Industry reports from Scientific Counsellors, no. 2, February 1972).
4. M. E. Burt, *Roads and the environment* (TRRL Report LR 441).
5. Report of Clean Air Society, cited in SMMT *Newsletter*, CVS/8 September 1972.
6. D. Lyon, J. M. Tims and K. Muller (Esso Research Centre), 'Techniques for reducing exhaust emissions from diesel engines', paper given at 14th International Congress of FISTA, London, June 1972.
7. L. H. Watkins, *Urban transport and environmental pollution* (TRRL Report LR 455, 1972).
8. D. M. Colwill, *Atmospheric pollution from vehicle emissions:* measurements in *Reading 1971* (TRRL Report LR 541, 1973).
9. M. Jacobson, *Air Pollution and the Motor Vehicle* (Automobile Association Memorandum, Oct. 1971).
10. G. McConnell, 'Fuel and combustion aspects of pollution control with diesels', paper given to Automobile Engineers' Conference on Noise, Dirt and the Diesel, March 1972.
11. D. M. Colwill and A. J. Hickman, *The concentration of volatile and particulate lead compounds in the atmosphere: measurements at four road sites* (TRRL Report LR 545, 1973).
12. D. F. Bampton, 'Legislation of noise and exhaust emission and the effect on commercial vehicle engines', paper given to the National Conference of the Institute of Road Transport Engineers, May 1972.
13. *Effects of traffic and roads on the environment in urban areas* (OFCD Report, July 1973).
14. *Environmental Bulletin*, Department of Industry, no. 27 (May 1974).
15. *Environmental Bulletin*, Department of Industry, no. 28 (June 1974).
16. J. Dickson-Simpson, 'Making diesels sociable', *Motor Transport*, January 12 1973.
17. C. Haynes and R. Lindsay, 'Comparative study of methods of reducing exhaust emissions to very low levels', paper given at 14th International Congress of FISTA, London, June 1972.

## Chapter five


1. A. A. Walters, *The Economics of Road User Charges* (WBSOP, no. 5), 1968.
2. J. R. Meyer, Peck, Stanason and Zwick, *The Economics of Competition in the Transportation Industries* (Cambridge, Mass., 1959).
3. *The Road Track Cost Report* (HMSO, 1968).
4. *The cost of constructing and maintaining flexible and concrete pavements over 50 years* (RRL Report, LR 256, 1969).
5. 'Transport policy – roads and road finance', *Socialist Commentary*, April 1975.
6. Jennings, T., 'The benefits of vehicle legislation; measuring the unmeasurable', *Transportation, 4.* no. 2. (1975).
7. Evidence by the Ministry of Transport to the Committee on Carriers Licensing on the Measurement and Allocation of Road Costs, 1964.
8. *Roads in England, Year ended 31st March, 1973* (HMSO).
9. International Road Transport Union, *Study and Research on Road Technique and the Economy of Infrastructures* (1965).
10. Road Research Laboratory, *A guide to the structural design of pavements for new roads* (Road Note, 29; HMSO, 1970).
11. A. Harrison, 'Some notes on road transport costs (University of Oxford Institute of Economics and Statistics, Bulletin, 27 no. 2, 1965).
12. A. A. Walters, 'The long and short of transport' (University of Oxford Institute of Economics and Statistics, Bulletin, 27 no. 2, 1965).
13. B. T. Bayliss and S. L. Edwards, *Industrial Demand for Transport* (HMSO, 1970).
14. R. Rees, 'Second-best rules for public enterprise pricing', *Economics* (1968).
15. C. D. Foster, *The Transport Problem* (1975).
16. D. Croney and L. H. Watkins, 'Heavy vehicles – research aspects', Symposium on the Heavy Vehicle, Institute of Highway Engineers, 1974.
17. *Offences Relating to Motor Vehicles* (HMSO, 1973).
18. D. Munby, 'Mrs Castle's Transport Act', *Journal of Transport Economics and Policy, 2* no. 3 (1968).
19. British Railways Board, 'A Study of the Relative True Costs of Rail and Road Freight Transport over Trunk Routes', memorandum to the Geddes Committee, 1964.
20. Ministry of Transport, 'The Measurement and Allocation of Road Costs', memorandum to the Geddes Committee, 1964, reproduced as Annex 2 in the *Road Track Cost Report.*
21. O. E. Williamson, 'Peak-load pricing and optional capacity under indivisibility constraints', *American Economic Review* (1966).
22. W. J. Baumol and O. F. Bradford, 'Optimal departures from marginal cost pricing', *American Economic Review* (1970).
23. K. Gwilliam *et al., Co-ordination of investments in transport infrastructures,* Transport Series, 3 (Brussels, 1973).
24. *Road Expenditure in Great Britain* (HMSO, 1974).
25. British Railways Report and Annual Accounts, 1974.
26. Department of the Environment, *The Continuing Road Goods Survey* (HMSO).


## Chapter six


1. *Road Casualties in Great Britain, Year 1974* (Department of the Environment. 1975).
2. *Road Accidents Great Britain, 1973* (Department of the Environment, Scottish Development Department, Welsh Office, 1974).

3. *Heavy Lorries*, A Civic Trust Report (October, 1970).
4. H. D. Johnson and F. Garwood, *Notes on road accident statistics* (RRL Report LR 394).
5. *Contributions to the accident situation by all types of vehicle* (TRRL Leaflet LF 320; reproduced in *Towards safer road vehicles*, TRRL Report LR 481).
6. H. P. Ruffell Smith, *A study of fatal injuries in vehicle collisions based on Coroners' reports* (RRL Report LR 316, 1970).
7. M. Miller, 'Accidents in which vehicles run under the rear of a commercial vehicle', unpublished but cited in Ruffell Smith, *op. cit.*
8. Independent Commission on Transport, *Changing Directions* (1974).
9. British Railways Board, Annual Report and Accounts, 1974.
10. *Survey of the Transport of Goods by Road, 1967–68* (Department of the Environment, 1971).
11. J. Ribat, 'Human factors and transport', Sixth International Symposium on Theory and Practice in Transport Economics, Madrid 1975, ECMT.
12. R. Kemp, I. Neilson, G. Staughton, and H. Wilkins, *A preliminary report on an on-the-spot survey of accidents* (TRRL Report LR 434).
13. M. Beesley and T. C. Evans, 'The costs and benefits of road safety measures'. ECMT Ninth Round Table (Organisation for Economic Co-operation and Development, Paris and Washington, 1971).
14. R. F. F. Dawson, *Costs of road accidents in Great Britain* (RRL Report LR 79, 1967).
15. M. Jones-Lee, 'Valuation of reduction in probability of death by road accident', *Journal of Transport Economics and Policy* (January 1969).
16. C. H. Sharp, *Transport Economics* (1973).
17. R. F. F. Dawson, *Current costs of road accidents in Great Britain* (RRL Report LR 396, 1971).
18. Transport Policies and Programme, 1971–7 (Warwickshire County Council, 1975).
19. Transport Policies and Programme, 1976–7 (Nottinghamshire County Council, 1975).
20. Transport Policies and Programme, 1976–7 (Leicestershire County Council, 1975).

## Chapter seven

1. 'Proposed Increases in Road Vehicle Weight and Size Limits', RIBA Memorandum, March 1970.
2. A. C. Whiffin and D. R. Leonard, *A survey of traffic-induced vibrations* (RRL Report LR 418).
3. *Heavy Lorries*, A Civic Trust Report (October 1970).
4. C. Sharp and A. Jennings, 'More powerful engines for lorries?', *Journal of Transport Economics* (May 1972).
5. *Guardian*, 31 January 1973.
6. A. Lassiere and P. Bowers, 'Studies on the social costs of urban road transport (noise and pollution)', ECMT Eighteenth Round Table, 1972.

## Chapter eight

1. A. J. Harrison, *The Economics of Transport Appraisal* (1974).
2. *Motorways and Trunk Roads*, Sixth Report from the Estimates Committee, 1968–9 (HMSO, 1969).

3. *Highway Statistics* (Department of the Environment, 1971).
4. A. J. Harrison and D. A. Quarmby, 'Theoretical and practical research on an estimation of time-saving, ECMT Sixth Round Table, 1969.
5. *New Roads in Towns*, Urban Motorways Committee Report, October 1972.
6. J. S. Rouse, 'Road expenditure decision-making at the local government level', Public Sector Economic Research Centre Paper, University of Leicester, 1975.
7. *Development and Compensation – Putting People First* (1973).
8. Land Compensation Act, 1973, Eliz.II, Chap. 26.
9. *Getting the Best Roads for our money: The COBA Method of Appraisal* (Department of the Environment, 1972).
10. Leicestershire Transport Policies and Programme, 1976–7.
11. Transport Policies and Programme, 1976–7, Warwickshire County Council.
12. Transport Supplementary Grant, Transport Policies and Programme, Submission for 1976–7, Nottinghamshire County Council.
13. Transport Policies and Programme, Submission for 1976–7, Kent County Council, July 1975,
14. W. G. Hoskins, *English Landscapes* (*BBC*, 1973).
15. *Heavy Lorries*, A Civic Trust Report (1970).
16. Clifford Chapman, 'Anatomy of the reviving giant', *Investors Chronicle*, 22 June 1973.
17. C. Buchanan, *Traffic in Towns* (1963).
18. Greater London Development Plan, Report of the Panel of Inquiry, 1973.
19. Tony Aldous, 'London's far-flung by-pass', and 'A choice of disasters', *Country Life*, 16 and 23 October 1975.
20. Interview with Dr John Gilbert, *Commercial Motor*, 26 September 1975.
21. Transport Policies and Programme, 1976–7, Lancashire County Council.
22. Transport Policies and Programme, 1976–7, Derbyshire County Council.
23. Transport Policies and Programme, 1976–7, Staffordshire County Council.

## Chapter nine

1. Independent Commission on Transport, *Changing Directions* (1974).
2. *Guardian*, 20 March 1972.
3. *Sunday Times*, July 1972.
4. *The Transport of Freight* (Ministry of Transport, 1967).
5. *Heavy Lorries*, A Civic Trust Report (1970).
6. *Roads and the Landscape* (Council for the Protection of Rural England, 1971).
7. Annual Report, British Railways, 1971.
8. Clifford Sharp, *The Allocation of Freight Traffic – A Survey* (Ministry of Transport, March 1970).
9. *The Reshaping of British Railways*, Part 1 (British Railways Board, 1963).
10. Statement by Minister of Transport in House of Commons, Sited Roadway, September 1975.
11. D. Aldcroft, *British Railways in Transition* (1968).
12. *Carriers Licensing*, Report of Geddes Committee (HMSO, 1965).
13. Minutes of Evidence, 1 March 1972, Expenditure Committee (Environment and Home Office Sub-Committee) inquiry on Urban Transport Planning.
14. Annual Report and Accounts 1974, British Railways Board, 1975.
15. *Roads in England 1972–73* (HMSO, 1974).
16. *Census of Production*, 1968.
17. *Census of Distribution*, 1971.

18. J. G. Walker, 'The description of railway noise', paper given to the Sixth Universities Transport Study Group Conference, Manchester, January 1974.
19. J. G. Walker, 'Noise from high speed railway operations', University of Southampton Institute of Sound and Vibration Research, Technical Report 75, May 1975.
20. John R. Kellett, *The Impact of Railways on Victorian Cities* (1969).
21. B. Bayliss, 'Raw material resources and transport', E.C.M.T. Sixth International Symposium on Theory and Practice in Transport Economics, Madrid 1975.
22. Estimated from U.N. Petroleum Industry Statistics, 1971 and 1972.
23. Estimated from *Energy Trends*, August 1975, Department of Energy.
24. P. Hatry, 'Raw material resources and transport', E.C.M.T. Sixth International Symposium on Theory and Practice in Transport Economics, Madrid 1975.
25. K. M. Gwilliam and P. J. Mackie, *Economics and Transport Policy*, (1975).
26. Information received from Professor Cheeseman in January 1974.
27. *The Energy Crisis and the Road Haulage Industry's Rates of Fuel Consumption* (International Road Union, Geneva, 1974).
28. British Railways Board, Annual Report and Accounts, 1970.
29. British Railways Board, Annual Report and Accounts, 1974.
30. Railways Act, 1974. 1974 Chapter 48, HMSO.

## Chapter ten

1. *National Travel Survey*, 1972–3 (Department of the Environment, HMSO 1975).
2. E. Dalby, *Pedestrians and shopping centre layout: a review of the current situation* (TRRL Report LR 577, 1973).
3. A. Jennings, C. Sharp and D. Whibley, *Delivering the goods*, F.T.A. Research Report no. 2 (1972).
4. 'Freight survey', *Economist*, 16 September 1974.
5. Chichester Central Area Servicing System, Local Interchange Depot Study (Nathaniel Lichfield and Partners, 1975).
6. P. W. Bryant, Chichester Town Centre, Interchange depot, An Environmental Assessment (West Sussex County Council, July 1975).
7. J. A. Battilane, *Compliance of heavy goods vehicles, with an entry restriction in Oxford* (TRRL Report 63 UC, 1974).
8. R. J. Bartlett, *Compliance of heavy goods vehicles with an entry restriction in Newbury* (TRRL Report SR 19 UC, 1973).

# Index

*Note* The only authors cited in the index are those mentioned in the text. The names of many other authors appear in the list of references and in the additional bibliography. Some widely used words such as 'car', 'bus', 'lorry', and 'train' have also been omitted from the index, and only selected major references to places and countries have been included.